Baedeker's

PRAGUE

914.3704
Bae
1991

Imprint

119 colour photographs, 17 maps and plans, 1 transport plan (Metro), 1 city plan

Conception:
Redaktionsbüro Harenberg, Schwerte

Text:
Dr František Kafka

Editorial work and continuation:
Baedeker Stuttgart

English language:
Alec Court

General direction:
Dr Peter Baumgarten, Baedeker Stuttgart

Cartography:
Ingenieurbüro für Kartographie Huber & Oberländer, Munich

English translation:
James Hogarth, Angela Saunders

Source of illustrations:
Doležal (1), dpa (12), Historia-Photo (12), Jürgens (1), Kusak (13), Sperber (1), Stoll (37), ZEFA (1)

Following the tradition established by Karl Baedeker in 1844, sights of particular interest are distinguished by either one or two asterisks.

To make it easier to locate the various places listed in the "A to Z" section of the Guide, their co-ordinates on the large city plan (and on the smaller inset plan of the city centre) are shown in red at the head of each entry.

Only a selection of hotels and restaurants can be given; no reflection is implied, therefore, on establishments not included.

In a time of rapid change it is difficult to ensure that all the information given is entirely accurate and up to date, and the possibility of error can never be entirely eliminated. Although the publishers can accept no responsibility for inaccuracies and omissions, they are always grateful for corrections and suggestions for improvement.

2nd Edition 1991

© Baedeker Stuttgart
Original German edition

© 1991 Jarrold and Sons Ltd English language edition worldwide

© 1991 The Automobile Association
United Kingdom and Ireland

US and Canadian edition Prentice Hall Press

Distributed in the United Kingdom by the Publishing Division of The Automobile Association, Fanum House, Basingstoke, Hampshire, RG21 2EA

The name *Baedeker* is a registered trademark
A CIP catalogue record for this book is available from the British Library.

Licensed user: Mairs Geographischer Verlag GmbH & Co.,
Ostfildern-Kemnat bei Stuttgart

Reproductions:
Golz Repro-Service GmbH, Ludwigsburg

Printed in Italy by G. Canale & C. S.p.A – Borgaro T.se – Turin

0–13–094806–3 US and Canada
0 7495 0285 1 UK

Contents

The Principal Sights at a Glance

Preface

This Pocket Guide to Prague is one of the new generation of Baedeker guides.

Baedeker pocket guides, illustrated throughout in colour, are designed to meet the needs of the modern traveller. They are quick and easy to consult, with the principal features of interest described in alphabetical order and practical details about location, opening times, etc., shown in the margin.

Each guide is divided into three parts. The first part gives a general account of the city, its history, notable personalities and so on; in the second part the principal sights are described; and the third part contains a variety of practical information designed to help visitors to find their way about and make the most of their stay.

The Baedeker pocket guides are noted for their concentration on essentials and their convenience of use. They contain numerous specially drawn plans and coloured illustrations. At the back of the book is a large plan of the city. Each entry in the main part of the guide gives the co-ordinates of the square on the plan in which the particular feature can be located. Users of this guide, therefore, will have no difficulty in finding what they want to see.

Facts and Figures

Prague's coat of arms

General

Czechoslovakia – officially the Czechoslovak Socialist Republic – is a federal union of two States, the Czech Socialist Republic (capital Prague) and the Slovak Socialist Republic (capital Bratislava). | **State**

The country's official languages are Czech and Slovak, which have equal status. | **Official languages**

Prague (in Czech Praha), the "hundred-towered" or "golden" city on the Vltava (Moldau), is capital of the Czechoslovak Socialist Republic (ČSSR) and of the federal Czech Socialist Republic (ČSR), and is also the seat of the administration of the Prague city district and the region of Central Bohemia. | **Capital**

Situated in a basin in the Vltava Valley in Central Bohemia, Prague lies 50° 05′ north and 14° 25′ east at the north-eastern tip of the Bohemian Silurian trough and on the edge of the chalk-belt. The Moldau or Vltava, a tributary of the Elbe/Labe, flows for 28 km (17 miles) through the city limits and is spanned by 15 bridges up to 300 m (330 yds) long. The bend in the river which is exposed on the western side has led to the formation of rounded slopes. So the broadly laid-out fortified town of Hradčany towers over a steep slope on the left bank of the Vltava, while the area containing the Old and New Towns on the right river-bank rises only gradually. The modern conurbation extends on all sides of the surrounding plateaux, Prague has an average altitude of 180 m (590 ft) at river level, rising to 383 m (1257 ft) in the White Mountain (Kopanina). | **Geographical situation**

The climate of Central Bohemia and Prague is influenced by their position in the centre of the northern temperate zone, at the junction of oceanic and continental climatic regions, and there is a clearly marked change in the seasons without variations in temperature being too severe (climate table see Practical Information: When to go). The monthly mean air temperatures are: January −0.9° C, July 19° C; the mean annual temperature is 9.0° C. The average annual rainfall is 487 mm/19 in. | **Climate**

In 1883 the old town of Prague had an area of 8·5 sq. km (3·3 sq. miles) and a population of 160,000. By 1901, after absorbing the neighbouring communes of Vyšehrad, Holešovice-Bubny and Libeň, it had increased in size to 21 sq. km (8·1 sq. miles), with a population of 216,000. In 1922 the local government entity of Greater Prague was established through the incorporation of 37 other districts, giving the city an area of 171 sq. km (66 sq. miles) and a population of 677,000. Prague now occupies an area of 497 sq. km (192 sq. miles) and has a population of about 1,195,000. In 1989 about 11,135,000 inhabitants occupied an area of 10,994 sq. km (4244 sq. miles) in the Central Bohemian region of Prague. | **Area and population**

◀ *Town Hall in the Old Town*

Population and Religion

City wards

Prague is divided for administrative purposes into ten wards. The first ward corresponds to the historic core of the city, taking in the Josefov quarter (the old Jewish quarter) and part of the Old Town (Staré Město), the New Town (Nové Město), the Malá Strana or Lesser Quarter and the Hradčany district.

Administration

The local government authority is the Central National Committee of 100 directly elected members, presided over by a *primator* (chairman). The ten wards have similarly constituted district committees.

Population and Religion

Population

In the 14th c., during the reign of Charles IV, Prague was one of the largest cities in Central Europe, with a population of some 60,000. The blood-letting of the Thirty Years War led to a sharp decline in population, and by the end of the 17th c. Prague had no more than 40,000 inhabitants. The population began to increase again during the 18th and 19th c. with the establishment of the city's metropolitan function, and further impetus was given by the industrial development of the later 19th c. At the turn of the century the population passed the 200,000 mark. Thereafter the city continued to grow at a rapid pace, which was still further intensified by the incorporation of adjoining communes. By 1930 Prague had 849,000 inhabitants. The upward trend was maintained after the Second World War, and in 1960 the population reached a million. Since the mid 1970s the demographic curve has visibly flattened, and recent years have seen only hesitant progress towards the 1·2 million mark.

Ethnic composition

Prague's largest population group has traditionally consisted of the descendants of the original Slav peoples, in particular the Czechs. In the 14th c., the period of Prague's rapid development, there was a substantial German minority; but by the 19th c. this had shrunk to something of the order of 5 per cent.
Over the centuries a considerable part in the city's life was played by the Jews, so that Prague came to be known as the "Jerusalem of Europe".

Religion

The majority religious belief in Prague has traditionally been Roman Catholicism. Second place is taken by the Czechoslovak Church (founded after 1918), which is Hussite. The largest Protestant community is the Slovak Evangelical Church of the Augsburg Confession. Recent estimates put the number of Jews at about 5000.

Transport

Port

The river harbour at Holešovice handles freight traffic, but is of no great commercial importance.
The Vltava is not navigable upstream from Prague, and even downstream is suitable only for vessels of modest size.
The landing-stage for passenger ships is on Engels Embankment (Nábřeží B. Engelse), near the Palacký Bridge.

Airport

Prague has a modern airport at Ruzyně, 20 km (12½ miles) north-east of the city centre. From Ruzyně the Czechoslovak national airline, ČSA, flies scheduled services to numerous cities through-

Main Station

out the world, and the airport is used by more than 25 international airlines.

Prague is linked with the European railway network, and its Central Station (Hlavní nádraží) in the Vitězného února is the terminus for numerous national and international services.
Other stations handle traffic within Czechoslovakia. There are also almost 30 small suburban stations for passenger traffic.

Rail services

Prague's Underground system (Metro) is still in the course of development. At present routes A, B and C are in operation and further stretches should be completed within the next few years.

Underground (Metro)

The trams are Prague's principal form of public transport, with services running on average at 15 minute intervals. Buses are mainly used to provide links with Underground stations and to serve outlying suburbs.

Trams and buses

Important main roads (federal highways, European highways):

Trunk roads

3 (E 14)	To Benešov, Votice, Tábor, České Budějovice and Linz
4	To Strakonice and Passau
5	To Plzeň (Pilsen) and Nuremberg (E 12)
6	To Nové Strašecí, Revničov and Karlovy Vary (Karlsbad)
7	To Slaný and Chomutov
8 (E 15)	To Lovosice, Teplice and Dresden
9	To Mělník, Děčín and Dresden
10 (E 14)	To Mladá Boleslav, Turnov and Jelena Gora
11 (E 12)	To Poděbrady, Hradec Králové and Wroclaw
12	To Kolín
12 (E 15)	To Brno

Culture

Prague has been for centuries one of the great cultural centres of Europe. Since the foundation in 1348 of the Charles University, the first university in Central Europe, contributions to the intellectual history of the West in the fields of learning, literature and art have been made here. In our own day Prague is the cultural centre of Czechoslovakia, with a large number of educational and research institutes covering every discipline, well-stocked libraries and record offices, laboratories and numerous museums. Apart from the collections on permanent display in the large art galleries, art-lovers will be impressed by the exhibitions staged at various galleries in the city.

Prague has 10 higher educational establishments, 26 adult art colleges, 34 technical colleges, 226 general schools and 416 kindergartens.

Theatre and Music

Prague has 20 permanent theatres including two opera-houses, three prominent philharmonic orchestras and many small musical ensembles well-known for their modern light music and jazz. Every kind of drama ranging from opera to comedy is performed at the variety theatres of which there are over 30. The "intimate theatres" (pantomimes, cabarets, puppet shows) and the combined film and stage shows of the "Laterna Magica" are particularly renowned for their originality. The highlights of musical and theatre life are the festival weeks (Prague Spring) held every year from the middle of May to the beginning of June, during which performances are given by world-famous orchestras and soloists from home and abroad. These take place not only in the concert halls but are also in some churches (St Vitus's Cathedral, St George's Basilica, St Nicholas's Church in the Lesser Quarter, Týn Church, St James's Church), museums, former mansions of the aristocracy and Baroque gardens.

Academies

In addition to the time-honoured Charles University, the Czechoslovak Academy of Sciences runs a considerable number of scientific institutes and research centres in Prague. The city also has an Academy of Art and a Conservatoire of Music.

Libraries and museums

Prague has several large libraries, more than 20 museums and numerous cultural organisations. Among the most important libraries are the National Library, the Municipal Public Library, the library of the National Museum and the National Medical Library.

Preservation of Ancient Monuments
Redevelopment of the city

There is scarcely another town in Europe with such architectural variety within such a small area as Prague. In the mid-eighties an extensive campaign was begun to restore the historical buildings in the centre of the city, and at the end of 1987 the Old Town Square was opened to the public as a splendid pedestrian zone. The blending pastel tones gave a new-look to the restored façades of the Gothic, Baroque, Renaissance and Rococo buildings in this charming square. Restoration of the interiors of the old houses was entrusted largely to private enterprise. The Town Hall of the Old Town, St Nicholas's Church and the Týn Church were also lovingly restored. Splendidly renovated façades in Neo-Latin and Baroque styles may be seen in the Celetna on the way to the Powder Tower. Prague Castle, also almost perfectly restored shines resplendently over the Lesser Quarter. Further restoration work is currently being carried out, for instance in the

The "House of Artists"

Karlovo, a stretch of the legendary, medieval royal way between the Old Town and Hradčany.
Whilst every endeavour is being made to preserve the historical quarter of the city, the outskirts are being left to go to rack and ruin. Neither financial means nor the necessary initiative exists for this once magnificent area to be saved.

Economy

New economic guidelines were issued in Czechoslovakia in 1984, but real structural changes were not indentifiable, and because of the strong ties within the RGW could only be carried out with the agreement of the USSR and other eastern block countries. The need for improvement to the economic mechanism once again led in the mid-eighties to consideration of economic reform which is to be put into action, following a preparatory phase (1988–1990), with the beginning of the 9th five-year plan (1991–1995). Among other things, the long-term reform provides for socialist autonomy within individual economic organisations and a stronger position in the world market. Strategic importance comes as a result of reform in the fields of electronics, automation, nuclear power – in the year 2000 50% of electronic energy will be provided by nuclear power – the manufacture of new products and bio-technics. Moreover, production of consumer goods as well as exports, is to be increased.

Economic Reform

Prague is the economic centre of Czechoslovakia, and all the important national economic organisations have their headquarters here. There are over 90 major industrial concerns in Prague, contributing almost 10 per cent of the total output of industrial

Industry

13

products. A key position has traditionally been occupied by the engineering industry; and other important branches of industry, in addition to metallurgy and metal-processing, are pharmaceuticals, foodstuffs, papermaking and textiles. The largest industrial regions lie in the north-east of the city (Libeň, Vysočany) and in the south-west (Smichov, Jinonice). The new laws outlined in 1989 will restrict central control of the economy and give more independence to national enterprises. There is provision for employees to elect their own management and for organisations to make business contacts direct with overseas companies. Part of the agreement comes into force in 1991.

Commerce

Long an important point of intersection of trade routes and the starting-point of the river-borne trade on the Vltava and the Elbe, Prague is still one of the leading commercial centres of Eastern Central Europe, and in recent years it has consolidated its reputation as a shopping centre, with more than a dozen large department stores and well over 4000 smaller shops offering a wide range of goods. Numerous banks and insurance companies have become established here.

Tourism

Prague is one of the most popular tourist cities in Eastern Europe, and according to the latest estimates more than a million visitors come to the "golden city" from the western neighbour countries every year.

Notable Personalities

The Danish astronomer Tycho Brahe, already famous when he came to Prague in 1597, became Court Astronomer to the Emperor Rudolf II in 1599. He had previously built an observatory in Denmark, and his astronomical instruments were the largest of their day (the telescope not having yet been invented).

His observations provided the empirical basis for Kepler's laws of planetary movement, but he himself always remained an opponent of the heliocentric picture of the Universe. The Tychonian system, retaining the earth as the centre of the Universe, long continued to compete with the Copernican system.

Brahe's memorial slab, with his portrait, is in the Týn Church (see A to Z).

Tycho Brahe
Danish astronomer
(1546–1601)

Peter Johann Brandl (or Brandel or Prantl), one of the great masters of Baroque art in Bohemia, was Court Artist to successive rulers of Bohemia and was also frequently employed by religious houses.

Subject to both Venetian and Flemish influences, he achieved a style of formal independence, based on his own vision and sensibility and showing marked realism. There are pictures by Brandl in a number of Prague churches.

Peter Johann Brandl
Bohemian painter
(1668–1735)

Max Brod made his name mainly as the editor of Franz Kafka's writings, which – against Kafka's express wishes – he published posthumously. He himself was an Expressionist writer, with works such as "Tycho Brahes Weg zu Gott" ("The Redemption of Tycho Brahe"), "Uber die Schönheit hässlicher Bilder" ("On the Beauty of Ugly Pictures"), "Streitbares Leben" (an autobiography) and "Die verkaufte Braut" ("The Bartered Bride"). He worked on the editorial staff of the "Prager Tagblatt" and was a member of the Prague Group. He wrote biographies of Kafka and the composer Janáček. In 1939 he emigrated to Tel Aviv.

Max Brod
Prague writer
(1884–1968)

After coming to Prague Christoph Dientzenhofer, one of a number of architects in this Bavarian family, worked nowhere else. Together with his talented son Kilian Ignaz he created the characteristic "Dientzenhofer Baroque", a synthesis of the old Bavarian system of pilasters and the baldacchino principle of the Italian architect Guarino Guarini, thus providing the basis for the last and finest phase of Central European Baroque church architecture. Buildings begun by the father in a somewhat conventional style were frequently carried to completion and stylistic perfection by the son.

Examples of this are Břevnov Abbey, the Church of the Nativity in Loreto and above all St Nicholas's Church in the Lesser Quarter, one of the most important churches of the Late Baroque period in Central Europe.

Among other buildings by Kilian Ignaz Dientzenhofer, now essential features not only of the townscape of Prague but of the whole Baroque cultural landscape of Bohemia, are the Villa Amerika and the churches of St John of Nepomuk on the Rock, St John of Nepomuk in the Hradčany and St Thomas. He also designed the Sylva-Taroucca Palace and the Kinsky Palace.

Christoph Dientzenhofer
(1655–1722)
Kilian Ignaz Dientzenhofer
(1689–1751)
German architects

With his numerous chamber works and his great symphonies Dvořák paved the way for the emergence of distinctively Slav music. During his three years in America as Director of the

Antonín Dvořák
Czech composer
(1841–1904)

15

National Conservatory in New York (1892–95) he also influenced many young American composers. This period is reflected in his most famous symphony, "From the New World".

Dvořák also achieved reputation as a composer of operas ("Rusalka", "The Jacobins").

Albert Einstein
German-American physicist
(1879–1955)

Albert Einstein, creator of the theory of relativity and the theory of gravitation, who received the Nobel Prize in 1921 for his contributions to quantum theory, was Professor of Theoretical Physics at the German University in Prague in 1911–12. There are commemorative plaques both on the building where he taught in the New Town (Viničná 1597/7) and on the house in which he lived in Prague-Smíchov (Lesnicka 1215/7).

Václav (Wenceslas) Hollar
Bohemian etcher and
draughtsman
(1607–77)

Václav or Wenceslas Hollar made his name with his etchings of Prague, London and various German towns. As a Protestant he was compelled to leave Bohemia in 1627, worked in Frankfurt as a pupil of Matthäus Merian and later in Strasburg and Cologne, travelled in Europe and North Africa and went to Britain, where he worked at the Court of Charles I. He died in London.

Hollar produced more than 3000 etchings and engravings. There is a large collection of his work in the Graphic Collection of the National Gallery in the Kinsky Palace (see A to Z).

Jan Hus
Czech reformer
(c. 1370–1415)

Jan Hus, Rector of the Charles University, preached in the Bethlehem Chapel against the authority of the Pope, criticised the secular possessions of the Church and called for a Bohemian national church. He was supported by the people and by King Wenceslas. In his "De ecclesia" ("On the Church") he set out his view of the Church as a non-hierarchical assembly of the faithful which acknowledged only Christ as its head, not the Pope. On the strength of a guarantee of protection by the German King Sigismund he appeared before the Council of Constance in 1415, accused of heresy. Refusing to retract his views, he was burned at the stake.

Hus's ideas formed the basis of the Bohemian Reformation, and his death set in train a vigorous revolutionary movement in Bohemia which eventually led to the Hussite Wars. Notable among the many memorials to the Reformer are the statues in Old Town Square and in the grand courtyard of the Carolinum. Adjoining the Bethlehem Chapel is a reconstruction of the house in which he lived.

Jan Hus

Karl IV

Johannes Kepler

John of Nepomuk took orders in 1370, and in 1380 was ordained as a priest. After studying law in Prague and Padua he was appointed Vicar-General of the archdiocese of Prague.

In 1393, on the orders of King Wenceslas IV, he was arrested, tortured and thrown in chains from the Charles Bridge into the Vltava: according to legend because he had refused to reveal to the King the secret of the Queen's confession, but according to the historians because he had appointed an abbot of Kladruby Abbey against the King's wishes. He was canonised in 1729 and became Patron Saint of Bohemia.

There are several statues of St John of Nepomuk in Prague. Perhaps the best known is the one on the Charles Bridge, erected in 1683 and soon followed by 29 other statues of saints.

John of Nepomuk is the best-known bridge-saint in Europe. He is invoked against danger from water and unjust suspicion. His normal attributes are a crown of five stars (representing the five lights which were said to have appeared over the spot where he was drowned), a crucifix, biretta and rochet.

*St John of Nepomuk
(c. 1350–1393)*

The writings of Franz Kafka, an insurance clerk by profession, attracted little attention in his lifetime. His novels ("America", "The Trial", "The Castle") and many of his short stories were published after his death by Max Brod, against his express wishes.

After the Second World War Kafka's work became the subject of international interest, and theologians and Communists, psychologists, philosophers and existentialists all sought their own interpretations of his very personal world. Two of his novels, "The Castle" and "The Trial", were made into films.

There is a bust of Kafka in Old Town Square. He is buried in the New Jewish Cemetery.

*Franz Kafka
Prague writer
(1883–1924)*

Charles became King of Bohemia in 1346 and Emperor in 1355. He founded Prague's University (the Carolinum) and built the Charles Bridge, St Vitus's Cathedral, the New Town of Prague and Karlštejn Castle, the last named as a place of safety in which to keep the Imperial insignia and Crown Jewels.

Charles made Prague the capital of the Empire, the "Rome of the North". During his reign poets such as Petrarch and Rienzo came to Prague and there was a great flowering of architecture. Charles himself was a writer, the author of the "Legend of St Wenceslas", the "Fürstenspiegel" ("Mirror of Princes") and the "Vita Caroli" (see Quotations).

*Charles IV
King of Bohemia and
Holy Roman Emperor
(1316–78)*

When Johannes Kepler was compelled to leave Graz during the Counter-Reformation he moved to Prague, where he succeeded Tycho Brahe as Court Astronomer to the Emperor Rudolf II in 1601. On the basis of Brahe's observations he formulated the laws governing the motion of the planets.

Kepler also did pioneering work in the field of optics and invented the astronomical telescope. After the death of Rudolf II he left Prague and moved to Linz.

*Johannes Kepler
German astronomer
(1571–1630)*

Egon Erwin Kisch achieved a considerable reputation as an active and vigorous reporter. His early years in Prague provided the material for works such as "Prague Adventures" and "Tales from Prague's Streets and Nights".

There is a commemorative plaque on the house in which he was born, U dvou zlatých medvědů ("At the Sign of the Two Golden Bears") in the Old Town (Kožná 475/1).

*Egon Erwin Kisch
Czech journalist and writer
(1885–1948)*

Notable Personalities

Josef Mánes
Czech painter
(1820–71)

Josef Mánes, originally influenced by the ideas of the Romantic school, is regarded as the founder of Czech landscape-painting and of a school of national folk-painting, to which he gave monumental scale. In 1848 he took an active part in the Czech Rising. His realistic paintings and illustrations (to folk-songs) depict country folk as ideal representatives of the people, and his cycle of the months on the astronomical clock in Old Town Square shows scenes of peasant life (originals in the staircase hall of the Municipal Museum).

There are pictures by Mánes in the Mánes Exhibition Hall and St Agnes's Convent. He is commemorated by a monument at the end of Mánes Bridge.

Wolfgang Amadeus Mozart
Austrian composer
(1756–91)

While the first performance of Mozart's "Figaro" in Vienna (1786) was a failure, its first performance in Prague was greeted with enthusiasm. "The good people of Prague understand me," declared the composer.

In October 1787 "Don Giovanni" was given its first performance in the Old Town Theatre (now the Tyl Theatre), and in 1791 Mozart wrote "La Clemenza di Tito" for the Emperor Leopold's coronation as King of Bohemia.

Mozart had many friends in Prague, and was on intimate terms with the music teacher F. X. Dušek and his wife Josephine. There is a Mozart Museum in their villa, the Bertramka, in Prague-Smíchov.

Alfons Mucha
Czech artist
(1860–1939)

The graphic artist and designer Alfons Mucha worked for a time as a scene-painter in Vienna, and then studied art in Munich and from 1888 in Paris, where he made a name for himself particularly with his posters for the actress Sarah Bernhardt. He also worked on interior decoration, applied art and book illustration. He was a major influence on the Jugendstil (Art Nouveau) movement. After spending some years in America (1904–10) he returned to Czechoslovakia. He is buried in Vyšehrad Castle.

František Palacký
Czech historian and politician
(1798–1876)

As a historian František Palacký saw the Hussite period as a central phase in Czech history. In 1848 he refused to take part in the German National Assembly in Frankfurt and presided over the Pan-Slav Congress in Prague, becoming a leading figure in the movement for the revival of Czech national feeling. There is a monument to him at the Palacký Bridge in the New Town.

Peter Parler
German architect and sculptor
(1330–99)

The German architect and sculptor Peter Parler played a major part in the development of Gothic art in Central Europe.

In 1353 he was summoned to Prague by Charles IV to continue work on St Vitus's Cathedral. He built All Saints Chapel in the Castle and designed the Charles Bridge, with the Old Town Bridge Tower, the sculpture on which also came from his workshop. His sons Wenzel and Johann carried on the construction of the cathedral. His nephew Heinrich worked in Prague as a sculptor.

Benedikt Ried von Piesting
Bohemian architect
(c. 1454–1534)

Benedikt Ried von Piesting was one of the principal representatives of Late Gothic architecture in Bohemia, the supreme exponent of the vaulted architecture of the Pre-Renaissance. He designed the Vladislav Hall in the Royal Palace in Hradčany Castle, one of the most magnificent secular buildings of its day.

Rainer Maria Rilke
Prague poet
(1875–1926)

Rilke was born in the Herrengasse in Prague. After a brief career as an officer, ended on health grounds, he studied art, philosophy and literature in Prague, Munich and Berlin. Many of his works were written under the influence of the Prague milieu of his day, a declining world of aristocratic and middle-class culture.

Egon Kisch

Rainer Maria Rilke

Bedřich Smetana

Smetana studied piano and musical theory in Prague and established his own music school there in 1848. After five years in Sweden he returned to his native city in 1861, and in 1866 became Conductor of the National Theatre Orchestra. He founded a Czech national style in both opera ("The Bartered Bride") and symphonic music ("Vltava"). Although he became deaf at the age of 50 he did not give up composition. He died in an asylum.
Smetana is buried in the cemetery in Vyšehrad Castle. There is a Smetana Museum at Novotného lávka 1 in the Old Town.

Bedřich Smetana
Czech composer
(1824–84)

Albrecht Wenzel Eusebius von Waldstein (Valdštejn), better known as Wallenstein, was one of the great generals of the Thirty Years' War. Accused of plotting high treason, he was dismissed and outlawed by the Emperor, and in 1634 was murdered together with his closest associates.
The Waldstein (Valdštejn) Palace in the Lesser Quarter was Prague's first Baroque palace.

Wallenstein
Imperial General in
Thirty Years' War
(1583–1634)

Wenceslas became Duke of Bohemia in 921. His grandmother and teacher, St Ludmilla, was killed by his mother Drahomira on religious grounds, and Wenceslas himself was murdered by his brother, Boleslav the Cruel, in 929 or 935.
Legend and reports of miracles made Wenceslas the country's national saint.
In Wenceslas Square, in front of the National Museum, is an equestrian statue of the saint with other national saints – Procopius, Ludmilla, Adalbert and Agnes. In the beautiful Wenceslas Chapel in Hradčany Castle can be seen a door-ring in the form of a lion's head to which Wenceslas is said to have clung when attacked by his brother.

St Wenceslas (Václav)
Duke of Bohemia
(c. 903–929 or 935)

Franz Werfel ranked with Brod, Kafka and Kisch as one of the great Prague writers of his day. He began by writing lyric poetry in the Expressionist manner and Symbolist dramas of ideas, but later turned to historical and political realism. His best-known works are the drama "Der Spiegelmensch" ("The Mirror Man"), "Der jüngste Tag" ("The Last Judgment"), a collection of poems, and the novels "Die vierzig Tage des Musa Dagh" ("The Forty Days of Musa Dagh"), "Die veruntreute Himmel" ("The Embezzled Heaven") and "Stern der Ungeborenen" ("Star of the Unborn").

Franz Werfel
Prague writer
(1890–1945)

19

History of Prague

4000 B.C.	Various tribes move from their heartland in Bohemia through the Vltava hills into the area now occupied by Prague.
3000–1000 B.C.	A trading settlement is established at the ford on the Vltava below the Hradčany, where the Amber road and the Salt road intersect.
From 400 B.C.	With the beginning of the Iron Age, the Boii, a Celtic people, move into Bohemia gradually subjugating the original inhabitants.
10 B.C.	The Boii are subjugated by the Marcomanni, a Germanic tribe.
6th c. A.D.	During the Great Migrations Western Slavs occupy the Prague area, with settlements on the castle hill and in what is now the Lesser Quarter.
800	Prague now consists of a number of fortified settlements. Legend has it that Prague was founded by Libuše, a princess with the gift of divination. According to her vision of a town whose fame is said to have once reached the stars, the princess's followers, as had been foreseen, established Prague on the Moldau, on the spot where a man was constructing the threshold (pràh) of his house. When the people tired of this feminine rule some years later, Libuše sent her henchmen to the Biela River. Near Staditz, as the princess had predicted, they encountered a young ploughman (přemysl) who was to become her future husband and first prince of Přemysliden.
c. 850–95	Duke Bořivoj, the first historically attested member of the Přemyslid dynasty, conquers the Czech tribes. The castle of Prague, the Hradčany, is built.
874	Bořivoj is baptised by Methodius, the Apostle of the Slavs. After his death his wife Ludmilla, also a Christian, is murdered. She becomes Bohemia's first martyr and patron saint.
921	Wenceslas, Ludmilla's grandson, becomes Duke of Bohemia.
929 (or 935)	Wenceslas is murdered by his brother Boleslav the Cruel (Boleslav I). After his canonisation as patron saint of Bohemia (Wenceslas' Glory) he becomes a lasting symbol of unity and independence in a land which often suffered under foreign rule.
973	During the reign of Boleslav II, the Pious, the episcopal see of Prague is established and the first St George's Convent is founded. The territory of Bohemia now extends to the boundaries of Kievan Russia.
900–1000	Jewish, German, Italian and French merchants settle in Prague.
993	St Adalbert, Bishop of Prague, founds the Benedictine Abbey of Břevnov.
1061–92	Duke Vradislav II (from 1085 King Vradislav I) transfers his residence from the Hradčany to the Vyšehrad.
1158	Duke Vladislav II is proclaimed King of Bohemia.

With the construction of the first stone bridge over the Vltava (later replaced by the Charles Bridge) Prague establishes its lasting dominance as a centre of trade.

Duke Soběslav II grants German merchants the right to be dealt with by German law, exemption from military service and fiscal privileges in order to encourage them to stay in Prague. **1178**

The Emperor raises the Duke of Bohemia (Přemysl Ottokar I) to the status of King who thus becomes the de facto ruler and from 1212 (Sicilian Golden Bull) inherits the title de jure. From 1289–1806 the Bohemian king also carries the title of Prince of the Holy Roman Empire. **1198**

Prague is fortified and receives its municipal charter. *c.* 1230

King Přemysl Ottokar II establishes the Lesser Quarter (Malá Strana) as a German settlement governed by the Magdeburg legal code.
In the following year, Ottokar extends his kingdom to include Austria and large territories in northern Italy, but is unsuccessful in his attempt to become Emperor. **1257**

The Prague groschen begins to be minted (63 groschen=1 mark). It circulates widely in Germany and becomes the model for the German groschen. *c.* 1300

With the murder of King Wenceslas III the Přemyslid dynasty dies out. **1306**

Following a period of unrest and, after a short reign, the death of Rudolf I in 1307, the Habsburgs make their first claim to the Bohemian throne. The German King Henry VII of the House of Luxemburg gives his son Johann in marriage to Elisabeth, heiress of the Přmyslid dynasty, and with support from the French and from the Church thus secures the throne of Bohemia for his family. **1310**

Charles IV rules Bohemia. Work begins on the construction of St Vitus's Cathedral, the archiepiscopal church of the newly created archdiocese of Prague. **1344**

Charles IV becomes King of Bohemia. **1346**

Charles becomes German King. He makes Bohemia the heartland of the Empire and unites Bohemia, Moravia and Silesia under the Bohemian Crown. As capital of the Holy Roman Empire Prague becomes the "Rome of the North". Scholars and artists flock to the city from all over Europe. **1347**

Foundation of the Charles University, the first university in Central Europe. The New Town is developed in grand style with huge squares (Wenceslas and Charles Squares), wide streets and churches and monasteries which blend well with their surroundings. There are also extensive fortifications which include the Gothic reconstruction of the Vyšehrad. The area of development is able to absorb the growing population, mainly craftsmen and tradesmen, for several centuries to come without the need for further expansion, and Prague becomes the largest city in Central Europe in both area and population. Charles builds the church of St Mary of the Snows and Karlštejn Castle. **1348**

Charles IV becomes Holy Roman Emperor. **1355**

Emperor Charles IV and his consort Anna

1357	Construction of the Charles Bridge and the Old Town Bridge Tower.
1378–1419	During the reign of Wenceslas IV there are severe social and religious tensions and conflicts over the throne. In 1400 Wenceslas is deposed as German King but remains King of Bohemia.
1409	On the urging of Jan Hus, Wenceslas curtails the rights of Germans in the universities 2000 German students and many professors leave the country and found various Universities including the University of Leipzig.
1415	The initial, peaceful attempts by the reformer, Jan Hus (See Famous People) and his many followers whose Christian teachings date back to its origins and whose aim it was to abolish differences within the ecclesiastical hierarchy, lead to a continual radical demand for religious, social and national reform. This becomes further inflamed when Hus, appearing before the Council of Constance, refuses to retract his views and is burned at the stake. His death triggers off a national anti-Church movement in Bohemia.
1419	On July 30, a mob led by Jan Žlivský storms the New Town Hall, frees the Hussites imprisoned there and throws two Catholic councillors out of the window. This first Defenestration of Prague signals the outbreak of the cruel Hussite Wars which are to last for 15 years. Death of King Wenceslas.
1420	Pope Martin V issues a Bull proclaiming a crusade against heretics in Bohemia. The Hussite army, led by Jan Žižka, defeats King Sigismund's crusading army in the Battle of Veitsburg.

Thereafter the Hussites, led by Prokop the Elder, take the offensive and mount retaliatory campaigns in Bavaria, Brandenburg, Saxony and Austria. Although they lose the war they gain some of their demands (expropriation of the Church's secular property, use of the chalice in Communion).

Following a short interregnum by Albrecht of Habsburg and after the throne had been vacant for 13 years, a Bohemian-Hussite nobleman, George of Poděbrad, first becomes regent and then king of Bohemia from 1458. Under his rule building activity continues in Prague (Týn Church).
The princes, enriched by the expropriation of Church property win increased influence. Prague's importance as a trading centre declines in favour of towns closer to the border. The importance of the University also wanes.

1436–71

The domains of the Bohemian Crown are united with Poland and Hungary. King Vladislav Jagiello transfers the capital from Prague to Buda.

1490

Following the death of Vladislav II's son Ludwig in the Turkish battle of Mohács, the Bohemian crown falls to his brother-in-law, Ferdinand I, a Habsburger. Extensive rights are thereby granted to the country and in particular to Prague (re-establishment of the Archbishopric, elevation to capital status). When restrictions come into force, however, a Prague led rebellion by the Towns and Estates against the king breaks out in 1547 and after the uprising is crushed the capital, Prague, and many other Bohemian towns are seriously punished by loss of privileges, rights and revenue.

1526

German Lutherans settle in Prague, reinforcing the opposition to the Counter-Reformation now introduced to Bohemia by the Roman Catholic Habsburgs.

From 1549

Ferdinand I becomes Emperor. He summons the Jesuits to Prague and building work gets swiftly under way. A new generation of strict Catholic noblemen and burghers emerges. This and the repeated attempts by the monarchy to curb religious freedom guaranteed by the Compact Acts of 1436, lay the foundations for continued struggles between the Bohemian Estates and the House of Habsburg also overshadowing the reign of Maximilian II (1576–1611).

1556

Maximilian's son, Rudolf II, lives in the Hradčany in order to pursue his interest in art collecting. The scholars, Tycho Brahe and Johannes Kepler (see Famous People) are appointed to assist him with his study of natural sciences and astronomy. He is attacked by his nephew, Leopold, and is compelled to appeal to his brother, Matthias and the Bohemian Estates for help. In return he concedes freedom of religion to the nobility in the 1609 "letter from his Majesty".

1609

Rudolf II abdicates and his brother Matthias becomes King.

1612

Renewed struggles over religious freedom and the privileges gained by the Towns and Estates lead to the second Defenstration of Prague on May 23rd. This is the signal for a rising by the radical Protestant nobility against the Catholic Habsburgs. Beginning of the Thirty Years' War.

1618

The Bohemian Estates depose the Habsburg Ferdinand II and choose Elector Frederick V of the Palatinate as King.

1619

1620   Frederick II defeats Frederick V, the "Winter King", in the Battle of the White Mountain (1619–1620) and establishes his hereditary right to the throne.

1621   Twenty-seven leaders of the rising of the nobility against the Habsburgs are executed in the Old Town on June 21st. The Protestant aristocracy and the wealthy middle-class are deprived of all power by banishment or the confiscation of property. Non-Catholics and their families are exiled.

1624   Ferdinand II moves the Bohemian Court Chancellery to Vienna, and Bohemia is ruled by officials responsible to Vienna. German and Czech remain the official languages, with equal status, but Czech gives place to German as a literary language.

1627   An Imperial Ordinance establishes the hereditary right of the House of Austria to the throne of Bohemia. Catholicism is the only permitted religion. The monarch has the pre-eminent right to legislate, appoint high dignitaries and annul resolutions of the Landtag (Diet). The power of the Estates is finally destroyed by this new constitutional law for Bohemia and Moravia, which remains in force until the 19th c.

1631   Wallenstein drives back the Swedes, who in the course of the Thirty Years' War have advanced to the gates of Prague.

1648   The news of the end of the war comes just as the Swedes are occupying the Lesser Quarter.
The Thirty Years' War has catastrophic effects on Bohemia. The country has lost almost half of its population and during the Wars which follow, heavy taxes are imposed by the House of Habsburg. Prague loses all cultural and economic importance.

Engraving of Prague in 1650

During the War of the Austrian Succession Prague is occupied by Bavarian, Saxon, French and Prussian forces.

1741–42

During the Seven Years' War Frederick the Great defeats the Austrians at Prague, but raises the siege of the city after his defeat at Kolín.

1757

Joseph II, continuing a process of reform begun in 1680, abolishes serfdom.
The use of the German language is further promoted.

1781

The Hradčany, Lesser Quarter, Old Town and New Town are combined to form a single unit.

1784

Opening of the railway between Prague and Vienna.

1845

A Czech national rising centred on Prague is crushed.
František Palacký refuses to take part in the German National Assembly in Frankfurt. The Pan-Slav Congress meets in Prague. Increased tensions between Germans and Czechs.

1848

Led by impassioned intellects and artists, the new Czechoslovakian Movement which emerged at the end of the 18th c. succeeds, after violent battles in Parliament, in repressing the use of the German language. The Germans lose their majority in the Prague Municipal Parliament for the first time.

1861

The peace which comes to Prague ends the War between Prussia and Austria for supremacy in Germany.

1866

Prague University is divided up according to nationality.

1882

A

M V L T A V I A   F L V V I V S

Historical development of the City of Prague

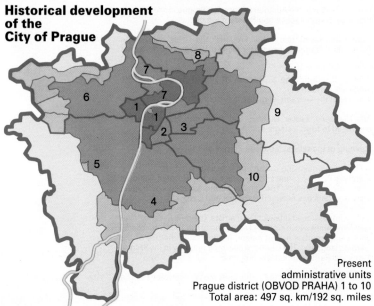

Present
administrative units
Prague district (OBVOD PRAHA) 1 to 10
Total area: 497 sq. km/192 sq. miles

Old town centre
Old Town, New Town, Lesser Quarter, Hradčany

Incorporations
1850: Josefov   1883–1901   1920/1922   1960   1968   1974

Administrative divisions of the City area

PRAGUE DISTRICT 1: Old town (Staré město), Josefov (Josefov), Lesser Quarter (Malá strana), Hradčany, Parts of New Town.

PRAGUE DISTRICT 2: New town (Nové město), Parts of vineyards (Vinohrady), Wyschehrad (Vyšehrad), Nusle-Tal (Nuselské údolí).

PRAGUE DISTRICT 3: Žižkov, Teile der Weinberge.

PRAGUE DISTRICT 4: Nusle, Michle, Spořilov, Krč, Lhotka, Podolí, Braník, Pankrác, Háje, Chodov, Kunratice, Libuš, Modřany, Hodkovičky, Cholupice, Písnice, Šeberov, Újezd u Průhonic.

PRAGUE DISTRICT 5: Smíchov, Košíře, Motol, Radlice, Jinonice, Hlubočepy, Malá Chuchle, Lahovice, Velká Chuchle, Lipence, Lochkov, Radotín, Řeporyje, Slivenec, Stodůlky, Třebenice, Zbraslav, Zličín.

PRAGUE DISTRICT 6: Bubeneč, Břevnov, Střešovice, Veleslavín, Vokovice, Dolní Liboc, Ruzyně, Dejvice, Sedlec, Řepy, Nebušice, Lysolaje, Suchdol, Přední Kopanina.

PRAGUE DISTRICT 7: Holešovice-Bubny, Letná, Troja.

PRAGUE DISTRICT 8: Karlín, Libeň, Bohnice, Čimice, Kobylisy, Stříškov, Dolní Chabry, Ďáblice, Březiněves.

PRAGUE DISTRICT 9: Vysočany, Prosek, Hloubětín, Hrdlořezy, Letňany, Čakovice, Kbely, Kyje, Běchovice, Dolní a Horní Počernice, Klánovice, Koloděje, Satalice, Újezd nad Lesy, Vinoř.

PRAGUE DISTRICT 10: Vršovice, Strašnice, Malešice, Záběhlice, Hostivař, Horní Měcholupy, Dolní Měcholupy, Petrovice, Štěrboholy, Benice, Dubeč, Kolovraty, Královice, Křeslice, Nedvězí, Uhříněves.

The German members of the Landtag withdraw. The movement of Czechs into towns of predominantely German population produces a change in ethnic structure, but the Germans still maintain their economic predominance.	1886

Industrial Exhibition in Prague. Industrialisation, particularly in areas of German settlement, has made Bohemia an industrial heartland of the Danube Monarchy. — 1891

Tensions between Germans and Czechs reduce the Landtag to impotence. During the First World War a state of emergency is declared in Bohemia. — 1913

Establishment of the Czechoslovak Republic, one of the Slav successor States to the Austro-Hungarian Monarchy. The first President is Tomás G. Masaryk.
The new multi-racial State is threatened by constant tensions between the various national groups (Czechs, Slovaks, Germans, Hungarians, Poles). — 1918

Prague's city limits are considerably extended by incorporation and 19 districts are formed. — 1922

Munich agreement: the German-settled territories of the Czechoslovak Republic are incorporated in Hitler's Germany. — 1938

The rest of Czechoslovakia becomes the Protectorate of Bohemia and Moravia, under Nazi rule. — 1939

After the end of the Second World War Zdeněk Fierlinger, a Social Democrat, proclaims the Košice Programme for a state on Socialist principles. — 1945

The Communist Party (KPČ) assumes power. Czechoslovakia becomes a People's Republic. — 1948

Prague is divided into 16 districts. — 1949

Establishment of the Czechoslovak Socialist Republic (ČSSR) and division of the city into 10 districts. — 1960

The "Prague Spring", under President Svoboda and First Secretary Dubček, is brought to an end by the intervention of Warsaw Pact troops. The Soviet Union gains the right to maintain troops in Czechoslovakia for an indefinite period.
In the same year, a further 21 suburbs come under municipal control. — 1968

Protesting against the entry of Warsaw Pact troops, a 20-year old philosophy student, Jan Palach, douses himself with petrol and burns to death in Wenceslas Square on January 16th. In a similar protest six weeks later, an 18-year old student Jan Zajic takes his life in the same manner. A new constitution establishes a federal state of two republics, the Czech republic (ČSR) and the Slovak republic (SSR) governed independently from Prague and Bratislava with a joint parliament within the Federal parliament in Prague. — 1969

On December 8th, a treaty is signed establishing inter-relations between Germany and Czechoslovakia. — 1973

The city is considerably enlarged by the incorporation of rural areas on the outskirts. The first Underground (Metro) line comes into operation. — 1974

27

History of Prague

1977	The Apostolic Administrator, František Tomášek is made a Cardinal by the Pope. A year later he is appointed Archbishop of Prague.
1981	The modern Palace of Culture is opened.
1984	West German Foreign Minister Genscher made a short visit to Prague in December.
1985	Gustav Husák (KPČ) re-elected State President in May.
1986	After 18 months, the Old Town Square is opened to the public, completely restored. Further projects should help to preserve the historical core of the City of Prague.
1987	Soviet leader Gorbachev made a friendly visit to Czechoslovakia in April.
1989	Thousands of anti-occupation demonstrators protest on August 21st, the 20th anniversary of the occupation by Warsaw Pact troops and the suppression of the 1968 Reform Movement. They campaign for freedom, civil rights and the rehabilitation of political activists of "Prague Spring". The demonstration through the Old Town is the largest in Czechoslovakia since 1969.

For 1989:

Thousands of anti-occupation demonstrators protest on August 21st, the 20th anniversary of the occupation by Warsaw Pact troops and the suppression of the 1968 Reform Movement. They campaign for freedom, civil rights and the rehabilitation of political activists of "Prague Spring". The demonstration through the Old Town is the largest in Czechoslovakia since 1969.

Demonstrations against the Government on October 28th in Wenceslas Square in memory of the establishment of the first republic 70 years ago are crushed by police using water-cannons and batons. Full scale action is used against known members of the Civil Rights Movement.

The entire Czechoslovakian Government resigns following the dismissal of the Prime Minister, Lubomir Strougal and the leader of the cabinet, Peter Colotka. (October). Ladislav Adamec becomes the new Prime Minister.

Police using force break up a human rights symposium in the Hotel Paříž attended by Prague civil rights supporters.

Several demonstrations marking the 20th anniversary of the day on which Jan Palach burned himself to death in Wenceslas Square are quelled by the police. Protest demonstrations by the Civil Rights Movement "Charta 77" are banned. (January).

In accordance with agreements by the Warsaw Pact countries, Czechoslovakia also agrees to reduce its armed forces.

The History of Art

First Sacral Building in the Greater Moravian Empire

The earliest remains of Christian stone churches found in the Czechoslovakia of today date back to the time when Konstantin (Cyril), the Apostle of the Slavs, and Methodius were missionaries in the Great Moravian Empire. There was a fortified settlement with five sacral buildings dating back to before the 9th c. A.D. in the south Moravian town of Mikulčice and in Staré Město u Uherského Hradiště (Old Town; possibly the "Veligrad" of Great Moravia) three churches were traced to the 9th century.

9th century

The Přemysliden Prince, Bořivof (c. 850–895) who was baptised at the Great Moravian court brought Christianity to Bohemia and during the first half of the 9th century founded the Church of St Clement at Levý Hradec on the banks of the Vltava north of present day Prague. After the prince's residence was moved to Prague, St Mary's Church, a small circular building, was constructed in 894. (The foundation walls are in the castle gallery.)

Romanesque (10th–12th c.)

St George's Basilica in the Hradčany was founded in 912 during the reign of Vratislav (c. 905–921). It was rebuilt between 1142 and 1150 (towers, east and west choirs, crypt) and to date represents the best preserved monument of Romanesque architecture in Prague.

A sacral building, significant because of its development, was the Rotunda of St Vitus, constructed between 926 and 930 by St Wenceslas. It was an Ottonian circular building with four apses and occupied the site on which the St Wenceslas chapel stands today in St Vitus's Cathedral. Several circular buildings with one nave (Bohemian rotundas) which had become characteristic of Bohemia followed later and other examples include the Holy Cross Chapel in the Old Town (c. 1100), St Martin's Chapel in the Vyšehrad (mid 11th c.; work probably began on this in the 10th c.) and St Longinus's Chapel (12th c.) in the New Town. St Vitus's Cathedral replaced the Rotunda of St Vitus in 1060. It has a double choir, a west transept and two crypts.

The remains of a Romanesque royal palace (9th–12th c.) built in Prague castle at about the same time are preserved beneath Vladislav Hall. In the second half of the 11th c. a Romaneseque stone castle and several churches stood in the Vyšehrad (remains of the St Lawrence Basilica).

Skilled crafts flourished in the convents founded from the end of the 10th c. (973 St George's Convent in the castle, 993 Brevnov, 1148 Strahov) and valuable illuminated manuscripts were produced in their scriptoriums. The most well-known of these is the Codex Vyssegradensis of 1086 which probably originated in Břevnov Abbey and which may be seen today in the Czechoslovak State Library in the Clementinum.

Gothic (13th c.)

To begin with, Gothic architecture was spread throughout Bohemia during the second quarter of the 13th c. by the Cistercians and orders of mendicant friars. The new style was soon reflected in secular buildings such as the Old-New Synagogue in Josefov (built by Cistercians in 1273) and the castle's Gothic palace (c. 1250–1400). It was also employed by the rising middle-classes on representational buildings such as the Town Hall of the Old Town (from 1338).

Late Gothic (14th c.)

During the reign of the art-lover and linguist, Charles IV (1346–1378) the countries of Bohemia began to look to the art form of

Central Europe. Influenced initially by the French, a change to the Late-Gothic style soon followed. The Royal Palace in Prague was consequently rebuilt, modelled on the French King's palace on the Ile de la Cité in Paris where Charles grew up. Summoned by Charles, Matthias of Arras from Avignon designed and re-built St Vitus's Cathedral on the same characteristic lines as the French cathedrals (choir with chapel cross). The foundation stone was laid on November 21st, 1344. After Matthias's death in 1352, Peter Parler (1330–1399) and his sons continued the work and gave the building a completely new, original look (with the emphasis on the south side – the Wenceslas Chapel, the main portal and tower). With his never tiring spirit of ingenuity, he brought to Prague a special style known as Parler Gothic which became a model for architecture and sculpture throughout Europe, reaching as far as Italy and Spain. Other cathedral choirs were created in the Parler workshops in Kolín and Kutná Hora. The Old Town's bridge tower in Prague which was not completed until the beginning of the 15th c. was also based on a design by Peter Parler.

Parler Gothic

Court art

Important early indications of the Renaissance period were featured in the creative art of the Charles IV era, the so-called court art: the first reticulated arches and portraits (the triforium busts in St Vitus's Cathedral) as well as the first free-standing equestrian statue (St George, by the brothers Martin and George Klausenburg).

Prague "mine of painters".

An original school of artists with varying styles developed when foreign (especially Italian) and certain Bohemian artistic trends were combined. The most prominent artists to emerge from the so-called "mine of painters" were the Master of Hohenfurt, active

St Martin's Chapel

St Vitus's Cathedral: Gothic interior

Old Town bridge tower: Charles IV – St Vitus – Wenceslas IV

in Prague around 1350, Master Theoderich (mentioned 1359–1380) and the Master of Wittingau (mentioned 1380–1390) who produced mainly altar-pieces (see A–Z, Hradčany, Collection of Old Bohemian Art in St George's Basilica; Karlštejn Castle).

The Byzantine influence is reflected in "Bohemian icons", half-portraits of the Madonna with the Christ Child painted by anonymous masters. The most significant frescoes, by unknown artists, are to be found in the Emmaus Abbey in Prague, founded in 1357 and in Karlštejn Castle (1348–1357) where Master Nicholas Wurmser (mentioned 1357–1360) from Strasburg and probably Tommaso da Modena (*c.* 1325–1379) worked. Splendid illuminated manuscripts were produced, often commissioned by the Silesian humanist, Johannes von Neumarkt who was responsible for the spread of German literary language in the State Chancery in Prague. An artistic decline set in following the deaths of Charles IV (1378) and Peter Parler (1399) but from this emerged, in contrast to the former realist style, the "beautiful" and "soft" style, a subtle, refined style of painting best known for the pictures of the Virgin which were produced around 1400.

Examples of Charles IV's designs and projects may still be seen today in many places in Prague (foundation of the New Town, Charles Bridge, Charles Court). He was unable to complete many of his ambitious plans during his life-time and with the outbreak of the Hussite Revolution (1419 – first Defenestration of Prague) which suddenly brought to an end the unprecedented golden age in the history of the city, some were never completed (Church of St Mary of the Snows) or were not completed for a long time (St Vitus's Cathedral, Týn Church).

Following the Hussite Wars, the character of Bohemian art – now almost purely Czechoslovakian – was, up to the end of the 15th c.,

15th c.

31

conservative and eclectic. The most famous Czech architect of that period was Matthias Rejsek (Powder Tower 1475). The most important artist summoned to Prague by King Vladislav II (1471–1516) was Benedikt Reid. He designed the Vladislav Hall in the Royal Palace in Prague Castle, one of the most splendid secular rooms of its period (1493–1502) and the main aisle in the church of St Barbara in Kutná Hora (1512–1547) with possibly the finest reticulated vaulting of the Late-Gothic period.

Just as many of the buildings are largely in the Renaissance style, sculpture and art of that period reflect almost all of the important schools and trends of the Dürer period. The most noteworthy among the indigenous artists is the Master of the Litomerice Altar (see A–Z, Hradčany, Collection of Old Bohemian Art in St George's Basilica) who around 1509 was engaged in the decoration of the Wenceslas Chapel in St Vitus's Cathedral, an excellent example of medieval wall-painting.

Renaissance (16th c.)

With the exception of Hungary, the art of the Italian Renaissance period probably became accepted earlier in Prague than elsewhere in central Europe. Based on designs by Paolo della Stella, Ferdinand I (1526–1564) had the arcaded building of Belvedere Palace (1538–1555) erected in the Royal Gardens of the castle. It was intended as a pleasure palace for his wife Anna and is one of the purest examples of Renaissance architecture north of the Alps. The original Star Castle was also built by Italians (1555–1558) based on an idea of Archduke Ferdinand. The ground plan is in the form of a star. After the middle of the 16th c. the leading architect in Bohemia was the master-builder to the Imperial court, Bonifaz Wohlmut from Überlingen on Lake Constance (organ gallery in St Vitus's Cathedral 1557–1561; Ballroom in the Royal Garden of the castle 1568; reticulated vaulting in the Hall of

Martinitz Palace

St Vitus's Cathedral: Wenceslas Chapel

the Diet in the castle 1559–1563; star-vaulted dome of the church of St Mary in Charles Court 1575). The new style now included features specific to Prague.

After the Great Fire of 1541 which destroyed an extensive part of the castle and the Lesser Quarter, residency of the aristocracy such as the Martinitz Palace (end of the 16th c.) sprang up in the Bohemian Renaissance style. The façades of these residences were frequently decorated with sgraffiti. This art form also often adorns Renaissance town houses such as the Three Ostriches House (1585 on the Charles Bridge or the House of the Minute (end of the 16th c.) adjoining the Town Hall of the Old Town.

Under the Habsburg King Rudolf II (1576–1611), a keen art collector, Prague became for the second time home to the Emperor and the centre of Mannerism. The emperor drew artists of every extraction to his court including the sculptors and bronze moulders Benedikt Wurzelbauer and Adriaen de Vries (who anticipated many Baroque elements), the artists, Hans of Aachen, Bartholomaus Spranger Jan Breughel ("Velvet-Breughel"), Guiseppe Acrimboldi, Roelandt Savery, Joseph Heintz, Hans Rottenhammer and Agidius Sadeler. The latter left behind a view of Prague in the year 1606 made up of nine engravings. It gives a graphic description of the architecture of that period and a reproduction of the work may be seen in the Castle Gallery.

Beyond the Court of Rudolf II Mannerism found no response in Bohemia. The etcher, Vaclar Hollar (1607–1677 see Notable Personalities) and the still life artist Gottfried Flegel (1563–1638) were both held in high esteem outside the country as representatives of Bohemian art of that period which was rather more firmly established. An example of the way in which Gothic archi-

Mannerism

Church of the Assumption

33

tecture had an influence on the Late Renaissance style may be seen in the Chapel of St Roch in Strahov Abbey.

Baroque (17th c.)

The final triumphant counter-reformation, following the victory by the Catholics in the Battle of the White Mountain (1621), brought in its early stages the predominantly Italian Baroque style to Prague. Mostly from the region of Como, the Italian craftsmen, affiliated to co-operatives, dominated the entire style of buildings for almost half a century. They built the Welsch Chapel for their own communities (1590–1600) which was the first important Baroque building in Prague.

The emphasis in the 17th c. lay firstly on the construction of palaces and castles. The most famous master builders among the Prague grandees were Albrecht von Waldstein (Wallenstein) who had a quarter of the town pulled down in order that his huge palace might be built (1623–1630) and Count Humprecht of Czernin (Czernin Palace, 1669–1692). Church buildings were at first modelled on the Jesuit 'Il Gesù' Church in Rome, the design of which was copied when alterations were carried out on the once Protestant Church of St Mary the Victorious (1611–1616). The most productive adepts of these Jesuit styles were Carlo Lurago (Church of St Ignatius 1665–1678) and Domenico Orsi de Orsini (the Court House in St Nicholas's Church in the Lesser Quarter 1673). Some church buildings, however, still continue to follow the older, local traditions (galleries, pilaster churches) as to some extent does St Salvator's Church on Knights of the Cross Square. Unlike architecture, sculpture and art was from now on dominated by local artists or by those who came from neighbouring countries. The most prominent artists after the middle of the 17th c. were Johann Georg Bendl (about 1630–1680; sculpture on the front of St Salvator's Church, Vintner's Hall on Knights of the

St Nicholas's Church in the Old Town

Cross Square, the statue of St Wenceslas outside the Old Provost's Lodging in the Castle) and Hieronymus Kohl (1632–1709; statues on the front of St Thomas's Church in the Lesser Quarter, fountains in the second courtyard of the castle). Among the painters, the most significant are Karel Skréta (1610–1674) who came from a family of Bohemian aristocrats and who emigrated because of his religious beliefs, later to return converted. He was the founder of Bohemian Baroque art. Also Michael Willmann (1630–1706) a pupil of the Rembrandt School (works by both artists are in the Collection of Old Bohemian Art in St George's Basilica.

The Late Baroque style reached its height in Prague during the first half of the 18th c., the most fruitful period for art since the time of Charles IV. The Frenchman Jean Baptiste Mathey (c. 1630–1695), who was trained in Rome, established a link with developments in Europe through church buildings – he built the Church of the Knights of the Cross (1679–1688) and through secular buildings – Troja Castle. Italian domination was now at an end.

Late Baroque (End of the 17th c. to the middle of the 18th c.)

The local artists were soon able to take the lead and after that the mark made on architecture by the Austrians and Bavarians of that time was assimilated. The Master Builder to the Viennese court, Johann Bernhard Fischer von Erlach built the Clam-Gallas Palace after 1707 and left behind several pupils in Prague. Christoph Dientzenhofer (1655–1722) a member of the extensive family of Bavarian architects moved to Prague, and together with his highly gifted son Kilian Ignaz Dientzenhofer (1689–1751) created the Dientzenhofer-Baroque style which combined both the old Bavarian pilaster system and the Baldachin principle of Guarino Guarini. With this, central European churches had finally reached their peak. Quite often buildings of a more conventional type began by his father were structurally completed by the son in a skilful, stylish manner. This is evident in Břevnov Abbey; in the Church of the Nativity in Loreto and quite particularly in St Nicholas's Church in the Lesser Quarter which represents one of the most important municipal church buildings in the history of the Late-Baroque period in central Europe. Christoph Dientzenhofer had already designed the dome chamber following the Bavarian pilaster principle. His son added the immense shell motif and with the non-uniformity of the building's external group of domes, the church stood out decidedly well on the Prague skyline. Further buildings by Kilian Ignaz Dientzenhofer, without which neither the character of Prague nor the remaining Baroque cultural landscape of Bohemia could be imagined are: the Villa Amerika, St Nicholas's Church in the Old Town, the churches of St John of Nepomuk in the Hradčany and St Thomas (remodelled) as well as the Ursuline Convent in Kutná Hora (Kuttenberg). The Kinsky Palace and Sylva-Taroucca which both have Rococo features were also designed by him. Giovanni Santin-Aichel (1667–1723) built the Morzin and Thun Palaces but both he and Oktavian Broggio (1668–1742) favoured churches built in an historic Baroque-Gothic style (St Mary's Church in Kuttenberg-Sedletz). Other architects of that period were František Maximilian Kaňka (1674–1766) who, among other works, designed the Vrtba-Terraces as well as various buildings contained within the Clementinum, and Giovanni Battista Alliprandi (1665–1720). Of the numerous sculptors active in Prague about 1700, the most renowned is Wenzel Matthäus Jäckel (1655–1738) whose works include statues on the Charles Bridge and in the former Church of the Knights of the Cross. Several groups of figures on the Bridge

Dienzenhofer Baroque

are by Ferdinand Maximilian Brokoff (1688–1731). The Berninesque style was brought to Bohemia by Bernhard Braun (1684–1738) and Bohemian Baroque sculptures which became famous were created in his workshops. Some of his statues may be seen on the Charles Bridge as well as in the Vrtba gardens and on the door of the Thun-Hohenstein Palace.

Rococo (2nd half of the 18th c.)

Rococo is represented in Prague by a number of artists and their works but, like Austria, Bohemia produced only a few original styles unlike Bavaria, Franconia and Potsdam. Prague is indebted to the architect and Knight, Count von Künigl, who planned the city's theatre (the present-day Týl Theatre built 1781–1783 by Anton Haffenecker) in which Mozart's "Don Giovanni" had its first performance in 1787.

The most prominent sculptors in the third quarter of the 18th c. in Prague were Johann Anton Quittainer (1709–1765), Ignaz Platzer the elder (1717–1787) – featured among his work is the sculptural decoration on the façade of the Archbishop's Palace – and Richard Prachner (1705–1782). Works by the last two artists may be seen in St Nicholas's Church in the Lesser Quarter. Known far beyond the borders of Bohemia was the subtle artist Norbert Grund (1717–1767), whose works even during his lifetime, were often forged or reproduced in the form of skesketches (See A–Z Hradčany, Collection of Old Bohemian Art in St George's Basilica.)

Classical

Of minor significance, at least as far as fine arts were concerned, was the Classical style. It was quite different from Baroque which had been more fervently adhered to in Bohemia than anywhere else in Central Europe. The empire façade (1808–1811) on Hibernian House, a former customs post and currently an exhibition

Plague Column

Archbishop's Palace

Hibernian House

Mánes Monument

hall, was worked by George Fisher and nevertheless one of the most noteworthy creations of this era.

The Romantic period was kindled in Bohemia when the Czech nation was awakened to the thought of Herder. The rise of the Czechs to cultural independence, however, led on the other hand to the decline of super-national "Bohemian Art" which from the end of the 19th c. had tended to become divided into separate Czech and German movements. Of the Romantic and Post-Romantic artists, the most prominent are Joseph von Führich (1800–1876), Joseph Mánes (1820–1871) and Mikuláš Aleš (1852–1913) and in addition Gabriel Max of the Piloty School and Václav Brožík.

Romantic

The most significant representatives of the Neo-Gothic style were Joseph Kranner (high-altar in St Vitus's Cathedral 1868–1873) and Joseph Mocker (extension to St Vitus's Cathedral 1859–1929). The Czech student of the Semper School, Joseph Zitek designed the impressive building of the National Theatre (1868–1881) in the style representative of Czech Neo-Renaissance. Following a great fire it had to be reconstructed by the architect, Josef Schulz (1840–1917) and was completed in 1883. The Artists' House "Rudolfinum", the finest buildings in Prague of this period, was also built by him.

Influenced by French art, the sculptor Josef Václav Myslbeck (1848–1922; equestrian statue of St Wenceslaus in Wenceslas' Square, bronze statue of Cardinal Schwarzenberg in St Vitis's Cathedral) founded a school from which emerged not only sculptors of the quality of Jan Štursa (1880–1925) but also Bohumil Kafka (1878–1942) among whose works is the Mánes memorial in front of the Artists' House, and Otto Gutfreund (1889–1927).

Josef Maria Olbrich (1867–1908), Josef Hoffman (1870–1956) and Adolf Loos (1870–1933), who did not however build in Prague but in Darmstadt (Olbrich), and Vienna were the forerunners of modern architecture in the 20th c. The switch to modern architecture in Prague was brought about by Josef Gočár (1880–1945) who between 1911 and 1912 erected the house, the "Black Mother of God" in the Celetna. The best example of the Prague "Secesson" style is the splendid House of Representation in the capital which was built between 1906 and 1911 and designed by Osvald Polivka and Antonin Balšanek. Conservative art trends are further maintained, however, certainly by the painter and illustrator, Max Šabinský (1873–1962) some of whose works include the magnificent glass paintings in St Vitus's Cathedral (1946–1948) and Heinrich Honer, a pupil of the Leibl School. Expressionism and Cubism were also represented, the first more among the German speakers (Oskar Kokoschka, Alfred Kubin, Josef Hegenbarth), the second among the Czechs (Emil Filla and Václav Spála). The monument in Old Town Square, dedicated in 1915 to Jan Hus, is a work by Ladislav Såloun who also created several allegoric sculptors for various buildings in the town. The art teacher and theorist Adolf Hölzel (1853–1934) of Olmütz is classed among the founders of abstract art.

Quotations

"Finally we came back to Bohemia after eleven years of absence. Our mother Elisabeth we found no longer alive: she had died some years before. And so we found, on our arrival in Bohemia, neither father nor mother, nor brother nor sisters, nor any other acquaintance.

"We had, too, quite forgotten the Bohemian language. Later, however, we recovered our command of it.

"This kingdom had fallen on evil days. No single castle was free; they had all been put in pawn, along with all the possessions of the crown, so that we had no place to stay, save in a house in the town like any other burgher. The castle of Prague had been so devastated, dilapidated and destroyed since the time of Ottokar that it had been wholly levelled to the ground. There we ordered the building, at heavy cost, of the spacious and stately palace.

"All honest Bohemians loved us, for they knew that we were a scion of the old royal house of Bohemia, and lent us their help in the recovery of the castles and the royal domains."

Charles IV
King of Bohemia
Holy Roman Emperor
(1316–78)
"Vita Caroli"

". . . the most beautiful inland town in Europe."

Wilhelm von Humboldt
German traveller and
scientist
(1769–1859)

"A first glance, then, reveals a Baroque city loaded with the spoils of the Austrian Caesars. It celebrates the Habsburg marriage-claims to the crown of Bohemia and reaffirms the questionable supersession of the old elective rights of the Bohemians; and alongside the Emperor's temporal ascendancy, this architecture symbolises the triumph of the Pope's Imperial champion over the Hussites and the Protestants. Some of the churches bear witness to the energy of the Jesuits. They are stone emblems of their fierce zeal in the religious conflict. . . .

"But in spite of this scene, a renewed scrutiny of the warren below reveals an earlier and a medieval city where squat towers jut. A russet-scaled labyrinth of late medieval roofs embeds the Baroque splendours. Barn-like slants of tiles open their rows of flat dormers like gills – a medieval ventilation device for the breeze to dry laundry after those rare washing-days. Robust buildings join each other over arcades that are stayed by the slant of heavy buttresses. Coloured houses erupt at street corners in the cupola-topped cylinders and octagons that I had first admired in Swabia, and the façades and the gables are decorated with pediments and scrolls and steps; teams of pargetted men and animals process solemnly round the walls; and giants in high relief look as though they are half immured and trying to elbow their way out. Hardly a street is untouched by religious blood-shed; every important square has been a ceremonious stage for beheadings.

"The symbolic carved chalices, erased from strongholds of the Utraquist sect of the Hussites – who claimed communion in both kinds for the laity – were replaced by the Virgin's statue after the re-establishment of Catholicism. Steel spikes, clustered about with minor spires, rise by the score from the belfries of the older churches and the steeples of the river barbicans, flattened into sharp wedges, are encased in metal scales and set about with spikes and balls and iron pennants. These are armourers' rather than masons' work. They look like engines meant to lame or

Patrick Leigh Fermor
English writer
(b. 1915)
"A Time of Gifts"
(referring to a journey in
1933)

hamstring infernal cavalry after dark. Streets rise abruptly; lanes turn the corners in fans of steps; and the cobbles are steep enough to bring down dray-horses and send toboggans out of control. . . .

"These spires and towers recalled the earlier Prague of the Wenceslases and the Ottokars and the race of the Přemysl kings, sprung from the fairy-tale marriage of a Czech princess with a plough-boy encountered on the banks of the river. . . ."

Fynes Moryson
English traveller
(1566–1630)
"An Itinerary", 1617

"On the West side of the Molda is the Emperours Castle, seated on a most high Mountaine, in the fall whereof is the Suburbe called Kleinseit, or little side. From this Suburbe to goe into the City, a long stone bridge is to be passed over Molda, which runnes from the South to the North, and divides the Suburbe from the City, to which as you goe, on the left side is the little City of the Jewes, compassed with wals, and before your eies towards the East, is the City called new Prage, both which Cities are compassed about with a third, called old Prage. So as Prage consists of three Cities, all compassed with wals, yet is nothing lesse then strong, and except the stinch of the streets drive back the Turkes, or they meet them in open field, there is small hope in the fortifications thereof."

Charles Sealsfield
(real name Karl Anton Postl)
Moravian writer
(1793–1864)

"All in all, Prague is one of the finest and most picturesque cities on the continent, much more interesting than Berlin or any other German capital. The extraordinary historical treasures of Prague make the city worth the closest observation. It would be a foolish enterprise to write a history of the world without previously visiting this ancient capital."

Adalbert Stifter
Austrian writer
(1805–68)
"Witiko" (a historical novel)

"There stood on a crag on the banks of the Moldau, before its waters reached Prague, the Castle of Vyšehrad. It was built when primeval forest still covered all these hills on the Moldau, long before the hero Zaboy lived and the singer Lumir. And then came Krok and had his golden residence in the sacred castle. Then there was Libuša, who among all her sisters was his favourite child, and she married the ploughman Přemysl and caused the first wooden stake for the Castle of Prague to be hewn. And from her there came numerous descendants, and they ruled over the peoples. One of them had himself baptised after Christ was born and brought the holy faith into the world. He was called Duke Bořivoj. His grandson was St Wenceslas and his wife St Ludmilla. He built the first church in Bohemia in his Castle of Hradec. Then at once he built the Church of the Blessed Virgin Mary in the Castle of Prague. In this church Duke Vladislav celebrated the cutting off of the hair of his son, St Wenceslas, and to this day it brings salvation to all believers. There, too, is the high Church of St Vitus. It was built with great labour and pains. St Wenceslas built it, and the Bishop of Regensburg, Tuto, granted him permission. Then Bishop Tuto died, and he who came after him, Bishop Michael, consecrated the church. It glowed with gold and silver and was full of splendour. And since it was too small Duke Spytihněv pulled it down and rebuilt it much larger, and then it was burned down and rebuilt again, and then lightning destroyed the tower and the tower was built anew. The most sacred treasures are in it. The German King gave St Wenceslas an arm of St Vitus for the church. Then the body of St Wenceslas himself was buried in it, and since that time many wonders have come about. And the body of the holy martyr Adalbert rests in it, and his vestments are preserved in its treasury, and the body of martyr Podiven, the faithful servant of St Wenceslas, is buried in it, and the body of Radim, the brother of St Adalbert. The church cannot accommodate the host of worshippers when it is the Feast of St Wenceslas and the sick come from foreign lands to be cured and when the Feast of St Adalbert is celebrated. This church is the most sacred church in the whole land of Bohemia. Then there is also the Church of St George. It was built still earlier than the Church of St Vitus. It was built by Duke Vladislav, son of the Duke Bořivoj who was baptised and father of St Wenceslas. Then he was buried in it, and the body of his mother, the holy martyr Ludmilla, also rests there. Beside it is the convent of the pious women of St George, where now the wounded are cared for."

Střešovická

HRADČANY

Jeleni

U Prašného mostu

Mariánské hradby

Královská zahrada

Belvedér

Na Opyši

Gogolova

Lete

Hanavský pavilón

Nový Svět

Keplerova

Arcibisk. palác

Katedrála sv. Víta

Valdštejnská

Vltava

Hote cont

náb

Pionýrů

Schwarzenberský palác

Loreta

Czernin palác

Myslbekova

Loretánská

Ke Hradu

Pražský hrad

Thun palác

Valdštejnský palác

Umělecko průmyslové muzeum

Mánesův most

Dům umělců

Ži m

Kaprov

Úvoz

Nerudova

Morzinský palác

Kostel sv. Mikuláše

Vojanovy sady

U tří pštrosů

sv. M

Vlašská

Mostecká

Karlův most

Klementinum

Strahovský klášter

Vlašská

Lobkovický palác

Vrtba palác

Kostel P. Maria Vítězné

Karmelitská

Harantova

Staroměstská mostecká věž

Karlova

M

Strahovská

Rozhledná

Tyrš dům

Smetanovo muzeum

Betlémská Kaple

Spartakiádní stadión

Spartakiádní

Spartakiádní

Kostel sv. Vavřince

Štefánikova hvězdárna

Kostel sv. Jana Nep.

ostrov

Rotunda svatého Kříž

Chaloupeckého

Šermiřská

MALÁ STRANA

Vítězná

most 1. máje

Národní div.

Ostrovn

Na Hřebenkách

Kinského zahrada

nám. Sovět. tankistů

Střelecký

Na Hřebenkách

Národní

Holečkova

Klova

Zborovská

Kampa

Smetanovo

Gottwaldovo

náb

Národopisné muzeum

Peškove

Mánes galerie

U Fleků

Novom r

Švédská

Sacré-Cœur

Viktora Huga

Myslíkova

Holečkova

Grafická

Kartouzská

Matoušova

Jiráskův most

Resslova

Koste ign

Plzeňská

Mozartova

Sv. Václav

Zborovská

Palackého most

Nábř. Engelse

Karlov

V něm Fau

Na Věnečku

Plzeňská

Lidická

Nádraží

Nábř. Engelse

Klášter na Slovanech

U Blaženky

Radlická

Kostel sv. Jana Nep. na Skalce

Bot zahr

Plzeň

Mrazovky

Na Zatlance

Ostrovského

Vltavská

Hotelsti Nábř.

Svornosti

Vltava

Nábr. B. Engelse

Apo

Kostel P. Marie u alžbětinek

U Nikolajky

SMÍCHOV

Na Václavce

Na Skalce

Radlická

Nádraží

Nábr.

Svobodova

Nektanova

Nábr.

Xaveriova

Xaveriova

Šantoškou

SMÍCHOV

Na pláni

VYŠEHRAD

Karla

Radlická

Radlická

Strakonická

Kostel sv. Petra a Pavla

Pechlátova

Pod Kesnerkou

Koulka

Praha-Smíchov Nádr.

Nádraží

Marxe

RADLICE

Metro

Krivoklát, Koněprusy,

Karlovy Vary, Lidice

Flughafen Ruzyně, Bila Hora

Sights from A to Z

Adria Building

See Cloth Hall

Archbishop's Palace

See Hradčany Square

Artists' House (Dům umělců)   D4

Artists' House is now the home of the Czech Philharmonic Orchestra (Česká Filharmonie). Concerts are given here, particularly during the internationally known Prague Spring Festival (May–June).

The Building, erected in 1876–84, was designed by Josef Zítek and Josef Schulz, the architects responsible for the National Theatre (see entry). It was originally called the Rudolfinum, after Crown Prince Rudolf of Austria, and is still familiarly referred to by that name.

Artists' House ranks with the National Theatre and the National Museum (see entry) as one of Prague's finest neo-Renaissance buildings. It was occupied from 1919 to 1939 by the Parliament of the Czechoslovaks and by a picture gallery which was later moved to the gallery in Hradčany Castle (see entry).

The decor in the Dvořák Hall is similar to that of the Gabriel Theatre in the Palace of Versailles. An allegory, dated 1885, of music by Wagner decorates the main entrance and the statues of the lion and the sphinx are the work of B. Schnirch. On the attic storey there are sculptures of famous artists and composers.

Just a short distance from the banks of the Vltava stands the statue, designed in 1951 by B. Kafka, of the Czech artist Josef Mánes, after whom the bridge which leads from here to the Lesser Quarter is named.

Location
Náměstí Krásnoarmějcú (Red Army Square), Staré Město, Praha 1

Metro
Staroměstská

Buses
133, 207

Tram
17, 18

Belvedere Palace (Královský letohrádek)   D3

This splendid pleasure palace was built by Ferdinand I for his wife Anna at the same time as he laid out the Royal Gardens (1538–63).

The arcaded building in the style of the Italian Renaissance with its gracefully curving roof was designed by Paolo della Stella. The upper floor was not completed until 1564.

The external colonnade is decorated with a frieze of foliage ornament and reliefs depicting scenes from Greek mythology and a likeness of Ferdinand I presenting a flower to his wife. The figures of divinities at the entrance are by Matthias Braun (c. 1730).

The Great Hall has frescoes by the Historical painter Christian Ruben (d. 1875). From the balcony there is a magnificent view of the Hradčany (see entry) and the city.

Location
Hradčany, Praha 1

Metro
Malostranská
Hradčanská

Tram
22

Opening times
Tues.–Sun. 10 a.m.–6 p.m.

◀ *St Agnes's Convent*

To the west of the Belvedere is the Singing Fountain (bronze, cast by Tomáš Jaroš in 1564).

To the south of the palace are the Chotek Gardens (Chotkovy sady). The Palace is closed at present for restoration.

Benatky nad Jizerou

Location
40 km (25 miles) north-east of Prague

Opening times
April– Sept. Tues.–Sun.
8 a.m.–3.30 p.m.

Situated on the banks of the Jizera (Isar), the town of Benatky nad Jizerou, founded in 1349, lies on the N10/E14 north-east of Prague. The town's splendid baroque castle has fine Renaissance sgraffiti and an interesting artesian well in the courtyard. The Benda family of musicians, the composer, Bedřich Smetana and the Danish astromoner, Tycho Brahe are the subjects of a fascinating exhibition in the castle museum.

Bertramka (Mozart Museum)   G2

Location
Mozartova 2/169, Smíchov, Praha 5

Metro
Anděl

Trams
4, 7, 9

The history of this residence began in the 17th c. when it was built by the Lesser Quarter master brewer Jan František Pimskorn. At the beginning of the 18th c. the villa came into the possession of František Bertram von Bertram after whom it was named.

From 1784–95 the suburban property was owned by the opera singer, Josefa Dušková, wife of the composer and music teacher František Xaver Dušek, and during his frequent visits to Prague Wolfgang Amadeus Mozart often stayed with the couple. In the mid 19th c. the Popelkas, father and son, founded the Mozart Memorial and a bust of the composer by the sculptor, Tomáš Seidan may be seen in the garden. Following extensive restora-

The Villa Bertramka

tion, the Mozart Museum, with exhibitions held in seven rooms on the first floor of the villa, was opened in 1956. Since 1976 the Smetana, the Dvořák and the Mozart Museums have been independent music departments of the National Museum. For the 200th anniversary of the first performance of Mozart's "Don Giovanni", Villa Bertramka was carefully renovated, right up to the end of 1987, and its collection was newly classified.

Wolfgang Amadeus Mozart (1756–91) visited Prague for the first time in 1787, following the successful performance in the town of his opera "The Marriage of Figaro". He made several visits to Prague after that, staying at the summer residence of his friends, the Dušeks. Among other concert pieces, he dedicated "Bella mia fiamma, addio" (KV 528) to the lady of the house. His famous opera, "Don Giovanni" was performed on October 29th, 1787 in what was then the Theatre of the Estates and is now the Týl Theatre. Commissioned four years later by the Impresario Guardasoni, Mozart composed the opera "La Clemenza di Tito" to mark the coronation of Leopold II when he became King of Bohemia. Mozart attended the première on his last visit to Prague shortly before his death in 1791.

The Bertramka now houses the Mozart Museum (original scores, Mozart's bedroom and study, exchange of letters, e.g. with G. Jacquin, historical placards, etc.). Concerts are also frequently given here. "Late Summer" is presented each September with way-out fashions and unusual props; in addition a show involving electronic music is planned.

Opening times
April–Sept.
Tues.–Fri. 2–5 p.m.
Sat. and Sun.
10 a.m.–noon, 2–5 p.m.
Oct.–March
Tues.–Fri. 1–4 p.m.
Sat. and Sun.
10 a.m.–noon, 1–4 p.m.

Bethlehem Chapel (Betlémská kaple)   E4

The Bethlehem Chapel was rebuilt in its original unpretentious form between 1950 and 1954 (architect Jaroslav Fragner) with the help of old descriptions and views. Declared a national monument in 1962, the sacral building is looked upon as one of the most significant religious monuments in Czechoslovakia.

In 1391 the burghers of Prague wanted to build a church in which the Mass would be said in Czech, but the Roman Catholic authorities agreed only to the construction of a chapel – though in the event it was a chapel which could accommodate a congregation of 3000 and was centred on the pulpit rather than the altar.

The Czech Reformer Jan Hus preached here between 1402 and 1413; he is commemorated by a memorial house adjoining the chapel. After his death the chapel remained the spiritual centre of the Hussite movement. In 1521 the German peasant leader Thomas Münzer proclaimed from the pulpit his vision of a communistic state and issued his Prague Manifesto.

From 1609–20 the chapel belonged to the Bohemian Brotherhood, one of whom was Jan Cyrillus who later became father-in-law to the great educator, Johann Amos Comenius.

After the Battle of the White Mountain (1621) in which Ferdinand II defeated the "Winter King", Frederick V of the Palatinate, and the Ordinance of 1627 declaring Catholicism the only permitted faith the chapel was acquired by the Jesuits.

In 1773 the Order was dissolved, and the chapel was pulled down in 1781 leaving only the foundations.

On the interior walls of the church may be seen fragments of treaties issued by Jan Hus and Jakoubek ze Stříbra. Students of the Academy of Graphic Arts have decorated the walls with paintings based on miniatures of the Jena Kodex, the Richenthalsch Chronicles and the Velislav Bible.

Location
Betlémské náměstí
(Bethlehem Square), Staré
Město, Praha 1

Metro
Staroměstská, Můstek

Trams
9, 17, 18, 21, 22

Opening times
Apr.–Sep. 9 a.m.–6 p.m.
Oct. 9 a.m.–5 p.m.

The Bethlehem Chapel

The reconstructed wooden pulpit, choir and oratorium are also of a more recent date.

Documents about the life and work of the reformer as well as the architectural history of the chapel are displayed in Jan Hus's memorial house.

Bilá Hora

See White Mountain

Botanic Garden (Botanická zahrada)   G4

Location
Ulice Na slupi,
Nové Město, Praha 1

The Botanic Garden has a tradition going back 600 years, and displays a corresponding variety of both native and exotic plants. The garden was laid out during the reign of Charles IV by a Florentine apothecary named Angelo. The University Garden, established in Smíchov in 1775, was transferred to this site in 1897.

Břevnov Abbey (Bývalý benediktinský klášter)

Location
Merkétská ulice,
Břevnov, Praha 6

5 km (3 miles) from the city centre on the Karlovy Vary road (No. 6) is the district of Břevnov, with a former Benedictine abbey founded by Prince Boreslav II and St Adalbert (Bishop of Prague) in 993 – the oldest monastery in Bohemia.

All that remains of the original Romanesque convent chapel which was built in the 11th c. is the crypt in the chancel. The Baroque complex standing today dates from the beginning of the 18th c.

Buses
108, 174, 180, 237

Trams
8, 22, 23

Courtyard
The courtyard is entered through a handsome gateway designed by Kilian Ignaz Dientzenhofer (1740).
In the courtyard can be seen a statue of St Benedict by Karl Josef Hiernle.

Conventual buildings
The Baroque conventual buildings were begun by Paul Ignaz Bayer in 1708 and completed about 1715 by Christoph Dientzenhofer. They now house the Central Archives of Prague. The fine Prelates' Hall has a ceiling-fresco by Cosmas Damian Asam (St Günther's miracle of the peacock, 1727).
The stucco work is by B. Spinetti and there are splendid paintings by A. Tuvora in the reception hall and the Chinese drawing-room.

St Margaret's Church
The central element in the abbey is St Margaret's Church (Kostel svaté Markéty), which was also built by Christoph Dientzenhofer (completed c. 1720). The ceiling-frescoes are by Johann Jakob Steinfels, the altar-pieces by Peter Brandl. The statue of St Margaret on the high altar is by Matthäus Wenzel Jäckel. In front of the church stands a statue of St John of Nepomuk by K. J. Hiernle.

Buquoy Palace (Buquoyský palác)   E3

The Buquoy Palace, now the French Embassy, was commissioned in 1719 by Marie Josefa von Thun (née Waldstein) and built to the design of J. B. Santini-Aichl, probably assisted by F. M. Kanka. It was extended in 1735. Sculpture work which adorns the palace is by M. B. Braun. After the palace was bought by the Buquoy family, the interior was decorated in neo-Baroque style. Designed by J. Schulz, the staircase and rear wing in neo-Renaissance style date from 1889–96 and further alterations were made in 1904. Two Gobelins (16th and 18th c.) which are worth seeing hang in the large hall in the rear wing. The old palace gardens, left in their natural state, extend down to Kampa Island (see entry).

Location
Velkopřevorské náměstí,
Malá Strana, Praha 1

Trams
12, 22

Carolinum (Karolinum)   E4

The Carolinum was founded by Charles IV in 1348 – the first university in Central Europe. The original nucleus of the building was the Röthlow House (1370), donated for the purpose by Wenceslas IV in 1383, the superb Gothic oriel of which is still preserved.
The architecture of the other university buildings ranges from Gothic to 20th century.
Soon, however, the history of the Carolinum as a university in the true sense, open to teachers and students from all over Europe, seemed to come to an end when, in 1409, Wenceslas gave way to the urging of the Reformer Jan Hus and curtailed the rights of Germans in the university, whereupon 2000 students and many professors left. Thereafter Hus (statue by J. Lidický in the Grand Courtyard) ruled the university as Rector, until in 1412 the

Location
Železná 9,
Staré Město, Praha 1
(Pedestrian zone)

Metro
Můstek

Gothic oriel on the Carolinum

Catholic faculty denounced him and he was compelled to flee to southern Bohemia. After the suppression of the rising by the Bohemian nobility the university was taken over by the Jesuits.

Of the original building there survive, in addition to the oriel, only the foundations and a few recently exposed Gothic features. The rest was remodelled in Baroque style by F. M. Kaňka in 1718.

The heart of the Carolinum, the oldest of its kind on the continent, is the large Assembly Hall (17th c.), two storeys high, which was extended in 1946–50 by Jaroslav Fragner.

*Celetná ulice (Bakers' Street)   E4/5

Location
Staré Město, Praha 1
(Pedestrian zone)

Metro
Staroměstská, Můstek

This street, named after the bakers who baked rolls (*calty*) here in the Middle Ages, was from time immemorial a route from the Vltava ford and the market-place of the Old Town to the east. This, too, was the route followed by the royal coronation procession from the Powder Tower (see entry) by way of Old Town Square and the Charles Bridge (see entries) to the Castle (see Hradčany). The street is lined for most of its length by handsome old palaces of Romanesque or Gothic origin, rebuilt in the Baroque period. One of the finest is the former Hrzán Palace (No. 12), which is believed to have been remodelled in High Baroque style by G. B. Alliprandi in 1702. The sculptures on the front of the house come most probably from the workshops of F. M. Brokoff. The Athenian figures on the Baroque Sixt House (No. 2), close to the Old Town Square are attributed to A. Braun. The family crest of the former owner of the Baroque palace, Caretto-Millesimo (No. 13) is above

House signs . . .   *. . . in the Celetná ulice*

the main door. The Spider wine-bar (U pavouka) moved to Menhart House (No. 17) whilst Buquoy Palace (No. 20) became the property of the Charles University in Prague following the alteration work carried out by A. Prachner in the 18th c. Prague High School has rooms in the neighbouring Vulture House (U zlatého supa, No. 22) and the much frequented Vulture Tavern may also be found there. The house was built in the Classical style. A statue of the Madonna (*c.* 1730) by M. B. Braun adorns the Baroque façade of U Schönfloku House (No. 83). The Pachtovský Palace (No. 31) was rebuilt in the Dientzenhofer Baroque style in the mid 18th c. and serves today as the Administrative Building. The much photographed house sign of the "Virgin behind bars" is the remaining evidence of a former Baroque façade on the House of the Black Mother of God (U černé Matky boží, No. 34) erected 1911–12. Where part of a Bohemian royal palace once stood in the 14th c. a mint has been in operation since the 16th c. The present mint (U Mincova, No. 36) was built by the Master of the Mint, František Josef Pachta of Rájova in the 18th c. A century later the building was extended and remodelled in neo-Baroque style. The monumental Powder Tower, a former Old Town fortification, stands at the end of Celetná ulice.

Český Šternberk (Bohemian Sternberg)

The mighty castle of Český Šternberk is one of the best preserved fortifications in Czechoslovakia. Because of its favourable position on the edge of an extensive stretch of woodland on a narrow rocky ledge overlooking the Sázava, the castle has, for a long time, remained impregnable. It was built as a private mansion

Location
45 km (28 miles) south-east of Prague

Charles Bridge

Opening times
Apr.–Oct. Sat., Sun.
9 a.m.–4 p.m.
May–Aug. Tues.–Sun.
8 a.m.–5 p.m.
Sept. Tues.–Sun.
9 a.m.–5 p.m.

during the reign of Wenceslas I (*c.* 1240) by Zdeslav of Divišov and Šternberk and with the exception of a short period during the 18th and 19th c. remained in the hands of the nobility until the mid 20th c. The scientist, Kaspar Maria Šternberk who was a co-founder of the National Museum in Prague (1818) was a member of the family.

During the reign of King George, the castle was captured and badly damaged. In 1479 repairs were made to it in Late Gothic style and it was further fortified with heavy bastions. After the Thirty Years' War, alterations to the castle were carried out in Early Baroque, the style which was popular in those days and much of the interior remaining today dates back to that time. The décor in the Great Hall, richly decorated Baroque fireplaces and the ornate stucco relief are also from the same period. Lovers of old-time music meet here during the summer months and the village of the same name at the foot of the castle is a popular holiday resort.

Charles Bridge (Karlův most) E3

Location
Staré Město, Praha 1
Karlův most

Metro
Staroměstská, Malostranská

Trams
12, 17, 18, 22

The Charles Bridge (closed to motor vehicles) spans the Vltava, linking the Old Town (Staré Město) and the Lesser Quarter (Malá Strana). From the bridge there are marvellous views of the Vltava Valley with its numerous bridges, the Žofin and Střelecký islands, the Old Town and the Lesser Quarter, dominated by Hradčany Castle (see entry). Under the western piers lies Kampa Island (see entry), separated from the Lesser Quarter by the Čertovka ("Little Venice"), a narrow arm of the river.

The bridge, supported on 16 piers, is 520 m (570 yd) long and 10 m (33 ft) wide. It was begun by Peter Parler and J. Ottl in 1357,

Charles Bridge – view of the Old Town Bridge Tower

Charles Bridge

Lesser Quarter Bridge Towers

Statues

→ **N**

St Wenceslas
by J. K. Böhm, 1858

SS. Cosmas and Damian
by J. O. Mayer, 1709

SS. John of Matha, Felix of Valois and Ivan and figure of a Turk
by F. M. Brokoff, 1714

St Vitus
by F. M. Brokoff, 1714 (marble)

St Adalbert
by F. M. Brokoff, 1709 (copy, 1973)

St Philip Benizi
by M. B. Mandl, 1714

St Luitgard
by M. B. Braun, 1710

St Cajetan
by F. M. Brokoff, 1709

St Nicholas of Tolentino
by J. F. Kohl, 1706 (copy, 1969)

St Augustine
by J. F. Kohl, 1708 (copy, 1974)

The **Charles Bridge** was begun in 1357 by Peter Parler, but completed only at the beginning of the 15th c. Its irregular course is probably due to the fact that after the collapse in 1342 of the first stone bridge over the Vltava (the Judith Bridge, built between 1158 and 1172), new piers were set beside the old ones but the original bridgeheads were re-used.

SS. Vincent Ferrer and Procopius
by F. M. Brokoff, 1712

St Jude Thaddaeus
by J. O. Mayer, 1708

Roland Column
(originally 16th c.; copy of 1884)

St Francis Seraphicus
by E. Max, 1855

St Anthony of Padua
by J. O. Mayer, 1707

St John of Nepomuk
by M. Rauchmüller and J. Brokoff, 1683; cast in bronze by W. H. Heroldt, Nuremberg, 1683

SS. Ludmilla and Wenceslas
workshop of M. B. Braun, c. 1730

SS. Wenceslas, Norbert and Sigismund
by J. Max, 1853

St Francis Borgia
by J. amd F. M. Brokoff, 1710 (restored by R. Vlach, 1937)

St John the Baptist
by J. Max, 1857

St Christopher
by E. Max, 1857

SS. Cyril and Methodius and three allegorical figures (Bohemia, Moravia and Slovakia), by K. Dvořák, 1938

St Francis Xavier
by F. M. Brokoff, 1711 (copy, 1913)

St Joseph
by J. Max, 1854

St Anne with the Virgin and Child
by M. W. Jäckel, 1707

Bronze Crucifix
cast by J. Hilger, 1629; the first piece of sculpture on the bridge, set up 1657; Hebrew inscription of 1696; figures by E. Max, 1861

Pietà
by E. Max, 1859 (originally 1695)

SS. Barbara, Margaret and Elizabeth
by F. M. Brokoff, 1707

Virgin with SS. Dominic and Thomas Aquinas
by M. W. Jäckel, 1709 (copy, 1961)

St Ivo
by M. B. Braun, 1711 (copy, 1908)

Virgin with St Bernard
by M. W. Jäckel, 1709 (copy)

Old Town Bridge Tower

during the reign of Charles IV, but not completed until the reign of Wenceslas IV, in the early 15th c. The massive towers at each end, like the bridge itself, were designed for defence.

Devastating floods have frequently damaged the bridge – in 1890, for example, two arches had to be rebuilt – but it has never collapsed.

"An avenue of statues"

The Charles Bridge achieves its powerful effect mainly through its rich sculptural decoration. This "avenue of statues", as it has been called, is mainly a product of the Baroque period; it is one of Prague's finest Baroque architectural compositions, and forms a remarkably effective combination with the severely Gothic structure of the bridge itself.

A bronze Crucifix which had stood here since the 14th c. was renewed in 1657, and between 1706 and 1714 26 pieces of sculpture by leading artists of the day (Matthias Bernhard Braun, Johann Brokoff and his sons Michael Josef and Ferdinand Maximilian) and other sculptors were set up on the bridge; these were followed in the mid 19th c. by five other pieces of sculpture (Josef Max and Emanuel Max); and in 1938 the group by Karel Dvořák representing St Cyril and St Methodius was added. The sandstone of the statues has suffered badly from the weather, and they are gradually being replaced by copies (The originals are in the Lapidarium in the National Museum). The only marble figure is that of St Philip Benizi.

The finest figure is that of St Luitgard, the very image of mercy and compassion. Christ is shown bending down from the cross towards the Saint and permitting her to kiss his wounds.

The only bronze statue is that of St John of Nepomuk, in the middle of the bridge, which was cast in Nuremberg in 1683 after models by Matthias and Rauchmüller and Johann Brokoff. Between the sixth and seventh piers of the bridge is a relief carving, marking the spot where St John of Nepomuk was thrown into the Vltava in 1393 on the order of Wenceslas IV because he had taken sides against the King in an ecclesiastical conflict. John of Nepomuk was canonised in 1729, and has been regarded since then as the "bridge-saint" of Catholic Europe.

On the Crucifixion group is a tablet with a Hebrew inscription, set up here by a Jew who was ordered by the court in 1696 to make this reparation for abusing Christ.

Old Town Bridge Tower

The Old Town Bridge Tower, built on the first of the piers, forms the eastern access to the bridge. It was begun in 1391 and completed in the early 15th c. (to the design of Peter Parler) by the Cathedral workshop. It is rated the finest Gothic tower in Central Europe, with figural decoration which ranks among the master works of Gothic sculpture in Bohemia (14th c.).

The Tower was renovated in the 19th c. by J. Mocker, when it received its present roof and original Gothic paintings restored by P. Maixner.

On the east side above the archway are the coats-of-arms of all the territories ruled by the House of Luxembourg as well as the royal Bohemian coat-of-arms consisting of the crest of the Roman Emperor and the royal kingfisher, the emblem of Wenceslas IV.

On the first storey there are statues of the enthroned Charles IV and Wenceslas IV with St Vitus between them. Above those there is a coat-of-arms bearing the eagle of St Wenceslas beneath a non-heraldic lion. At the top are figures of the patron saints of Bohemia, St Adalbert and St Sigismund.

The Virgin and St Bernard　　　　*St John of Nepomuk*

At the western end of the bridge are the Lesser Quarter Bridge Towers, linked by an arch. The lower tower (last quarter of 12th c.) formed part of the defences of the old Judith Bridge; the Renaissance pediment and the decoration of the outer walls were added in 1591. The other tower was built in 1464 at the behest of King George of Poděbrad on the site of an older Romanesque tower. Its architecture and sculptural decoration are similar to those of the Old Town Bridge Tower.

Lesser Quarter Bridge Towers

Charles Square (Karlovo náměstí)   F/G4

Charles Square is the largest square in Prague, measuring 530 m (580 yd) by 150 m (165 yd). Until 1848 the cattle market was held here. In its present form, with its lawns and trees and its statues of Czech scholars and writers, Charles Square is more like a park than a square.

On the south side of the square stands the Pharmacy of the Polyclinic, better known as the Faust House, on its east side the Church of St Ignatius, and at the north-east corner the tower of the former New Town Hall. In Resslova ulice, which leads from the west side of the square, are two other interesting churches, SS Cyril and Methodius and St Wenceslas of Zderaz (see entries). The neo-Renaissance Czech Technical College (No. 14) was built in 1867 to the design of V. I. Ullmann. Allegories depicting work and knowledge from the workshops of A. Popp flank the doorway whilst the sculptures of former scholars above the windows on the second floor are the work of J. V. Myslbek (1879).

Location
Nové Město, Praha 1
Karlovo náměstí

Metro
Karlovo náměstí

Trams
3, 4, 6, 16, 18, 21, 22, 24

House of the Golden Well

St Ignatius

Faust House (Faustův dům)

The Faust House, originally a palace of the Late Renaissance, was altered during the construction of the town's fortifications between 1606 and 1617 and had a corner bastion built on to it. In the 18th c. it was remodelled in the Baroque style. During the reign of Rudolf II (1575–1611) the English alchemist Edward Kelly carried out experiments here in the hope of producing gold. When another chemist installed his laboratory here in the 18th c. this gave rise to the legend that Dr Faust also occupied the house and, having sold his soul to the Devil, was carried off to Hell through the laboratory ceiling.

The Polyclinic thus seems to be following a well-established tradition in having its pharmacy here.

St Ignatius's Church (Kostel svatého Ignáce)

This Baroque church was built between 1665 and 1668 by Carlo Lurago, Architect to the Imperial Court, as the church of the former Jesuit college, which is now occupied by a section of the Polyclinic. The magnificent doorway (1697–99) (with a statue of St Ignatius in glory, 1671, on the front pediment) was the work of Paul Ignaz Bayer. The other statues are by A. Soldattis.

The sumptuous interior has rich stucco ornament by A. Soldatti. The 18th c. Baroque high altar made of marble depicts the "Glorification of St Ignatius Loyola" (1688), a work by J. G. Heintsch who is also responsible for several other altar pieces in the church. The painting of "Christ Imprisoned" is by his tutor Karl Škréta, and "St Liborius" was painted by Ignaz Raab. The picture beneath the organ-loft of figures on Calvary is by J. A. Quittainer.

New Town Hall (Novoměstská radnice)

This building at the north-east corner of Charles Square, originally Gothic, was erected about 1347 as the Town Hall of the New Town. After the amalgamation of the Hradčany, the Lesser Quarter, the Old Town and the New Town and the transfer of municipal functions to the Old Town (1784), the New Town Hall was used as a prison, lawcourt and registry office. In the corner tower, built in 1452–56 and subsequently much altered and rebuilt, is a chapel. At the beginning of the 16th c. the south wing was remodelled in the style of the Renaissance period; in the 19th c. empirical elements were added and in 1906 the original Renaissance style was finally reconstructed.

The New Town was occupied mainly by the poorer classes of the population. The First Defenestration of Prague took place here on 30 July 1419, when a mob led by the preacher Jan Želivský stormed the New Town Hall, freed the Hussites confined in the prison and threw two Catholic councillors out of the window, giving the signal for the beginning of the Hussite Wars.

Charles Street (Karlova ulice)   E4

This street forms part of the historic coronation route from the Old Town Square to Charles Bridge. The Baroque and Renaissance façades on the houses are at present undergoing renovation as part of the large-scale plan to restore the Old Town. The show-piece in the road is the "House of the Golden Fountain" (No. 3), restored between 1984 and 1986. The remains of masonry in the basement provide evidence that a Romanesque building once stood on this site. The Renaissance façade which may still be seen today has a splendid Baroque stucco relief by J. O. Mayer (1701), with pictures of SS Wenceslas, Roche, Sebastian Ignatius of Loyola, Francis Xavier and Rosalia. There is a cosy wine-bar, providing refreshments, on the ground floor. The Renaissance house "To the Golden Snake" (No. 18) is known for its ultra-modern interior and for being the oldest coffee-house in Prague. The Armenian D. Damajan who sold coffee in the streets of the Old Town and later opened Prague's first coffee-house in Three Ostriches House in the Lesser Quarter, lived here in 1714. The Disk Theatre, the experimental theatre of the State Conservatoire, is in Pötting Palace (No. 8) which has a Baroque façade and a crest in memory of its builder on the doorway. Prague's first cinema was installed in Blue Pike House (No. 20) in 1907 by Viktor Ponrepo and Dismas Šlambor.

Several shops in Charles Street sell Bohemian glass, pillow-lace and other forms of craftwork, while leading second-hand book-sellers including those in the Colloredo-Mansfield Palace (No. 2) and the "House of the Stone Mermaid" (No. 14) have a large number of volumes of rare publications for inspection.

Location
Praha 1, Staré Město,
Karlova ulice

Metro
Staroměstská
Můstek

Trams
17, 18

Charles University

See Carolinum

Clementinum (Klementinum)   E4

This former Jesuit college now houses the State Library of Czechoslovakia, with more than 5 million volumes, 6000

Church of the Assumption and Charles the Great

Location
Nám. primátora dr. V. Vacka
(Václav Vack Square), Staré
Město, Praha 1

Metro
Staroměstská

Trams
17, 18

manuscripts (including the Codex Vyssegradensis) and more than 4000 incunabula.

After inheriting Bohemia and Hungary through his wife in 1526 Ferdinand I sought to re-Catholicise his hereditary territories without resorting to unduly harsh measures. In 1556 he summoned the Jesuits to Prague, and they took over the Monastery and Church of St Clement (see entry), which had been occupied by the Domincans since 1232. A whole quarter of the Old Town – more than 30 houses, 3 churches and several gardens – was pulled down to make room for the Clementinum, the largest complex of buildings in Prague after the Hradčany (see entry). In 1622 the Charles University (see:Carolinum) was added to the complex.

In addition to five inner courtyards, the Clementinum includes within its area the churches of St Clement and St Salvator (see Knights of the Cross Square), a chapel belonging to the Italian community and an observatory (1751).

The main façade of the college building opposite the Church of St Francis was erected during the mid 17th c. It is richly decorated with shells, laurel, devil masks and busts of Roman emperors.

Notable features of the interior are, on the first floor, the Jesuit Library (known as the "Baroque Hall") designed by F. M. Kaňka with ceiling-frescoes of the Muses and biblical themes (closed to the public at the present time) and the Mozart Hall with Rococo paintings and book-cases from the same period. The former Mirror Chapel also built in 1724 by F. M. Kaňka has ceiling-paintings of the Madonna. It is now used for chamber concerts and exhibitions.

In the south-west courtyard can be seen a statue of a Prague student, commemorating the occasion at the end of the Thirty Years' War (1648) when the students defended the Charles Bridge against the Swedes.

Church of the Assumption and Charles the Great   G5
(Kostel Nanebevzetí Panny Marie a Karla Velikého)

Location
Nové Město, Praha 1
Ke Karlovu 1

Metro
Vyšehrad, I. P. Pavlova

Trams
6, 11

This church, with an octagonal plan modelled on Charlemagne's Palatine Chapel in Aachen, was built by Charles IV in 1358. It was consecrated in 1377, still with a temporary roof. In 1575 Bonifaz Wohlmut completed the stellar vaulting, one of the most remarkable achievements of medieval architecture, with a span of 22·75 m (75 ft). It is accounted for by the old legend of the builder who sold his soul to the Devil in order to be able to complete his master work. The Neo-Gothic high altar (1872) is the work of B. Wachsmann; the altar-pieces by A. Llotha and the ornate carvings by J. Brüllmayer and E. Veseláy. The pictures on the side altar of the Holy Family and of St Salvator were painted in 1708 by J. G. Heintsch. In the same year the statues in the chancel and chapel were presented to the church by J. J. Schlansovsky.

The church was remodelled in 1720 and subsequent years by Kilian Ignaz Dientzenhofer. The change in function from a monastic church to a place of pilgrimage from the early 18th c. and the addition of a number of chapels had the consequence of obscuring the original Gothic character of the church. Later the Baroque domes were added.

Museum of National Security

The Karlov Monastery, originally Gothic but remodelled in Baroque style by the Augustinian Canons in 1660–68, now houses the Museum of National Security and part of the State Archives. Art exhibitions are also occasionally held here.

Church of the Nativity

See Loreto

Cloth Hall (Dům Látek)   E4

The Cloth Hall, in the style of a Venetian Renaissance palazzo, was built in 1925–31 (architects, J. Zasch and P. Janák) for the Adria Insurance Company. It contains a piece of sculpture, "Adria", by Jan Štursa and other sculpture by Otto Gutfreund, Bohumil Kafka and K. Dvořák.

In the basement is the National Theatre's experimental theatre, the Laterna Magica. Special visual effects are produced by the combination of various techniques of theatre, music and film. The majority of these performances are given in the Palace of Culture (see Practical Information – Theatres).

Location
Národní 40, Nové Město,
Praha 1

Metro
Müstek

Community House (Obecni dům; Representační dům hlavního města Prahy)

See House of Representation for the capital of Prague.

Czernin Palace (Černinsky palác)   E1

This huge palace stands at the south-western corner of Loreto Square. Based on the design of Italian court architecture, its 150 m (492 ft) long façade was built in 1669 for Count Humprecht Johann Czernin of Chudenice, who was at that time the Imperial High Commissioner in Venice. His son Herman Czernin had the building completed in 1697. Italian builders and stone-masons were employed from beginning to end under the leadership of F. Caratti but after 1720 the palace was re-built by F. M. Kaňka. Added during this same period were the French gardens on the north-side and the magnificent stairway with a ceiling fresco ("Downfall of the Titans", 1718) by W. L. Reiner. The building was badly damaged during the French occupation of Prague in 1742 but was later repaired by Anselmo Lurago (1744–49). The orangery in the garden was rebuilt in Rococo style and three new front doorways were added. In the mid 18th c. I. F. Platzer produced several sculptures for the palace. Around the turn of the century, the building was taken over as a barracks, and reconstruction followed at the beginning of the 1930s. The palace now houses the Department of State for Foreign Affairs.

Location
Praha 1
Hradčany, Loretánské náměstí

Trams
22, 23

Dvořák Museum

See Villa Amerika

Emmaus Abbey/Abbey of the Slavs   G4
(Emauzy/Klášter na Slovanech)

This former Benedictine abbey (dissolved in 1949) now houses a number of institutes of the Czechoslovak Academy of Sciences.

Location
Vyšehradská, Nové Město,
Praha 1

Czernin Palace – seat of the Foreign Ministry

Emmaus Abbey

The abbey was founded by Charles IV in 1347, with Papal permission, for Benedictines of the Slav rite. Through this Order, reading the Mass in Old Slavonic, the Church sought to win over the eastern territories where the faith had made little headway.

In the 14th c. the abbey was an important centre of culture and education. The Glagolitic part of the Reims Gospel-Book, on which the kings of France swore their coronation oath, was preserved here until 1546. In 1945 the building was burnt out in an American air raid, and was not restored fully until 1967 when the original towers were replaced by two soaring, modern, concrete ones which resemble a pair of wings stretching up to heaven.

The frescoes in the Gothic cloister (much restored) are among the finest work of the old Prague school of painting. In the manner of the late medieval "Biblia pauperum", they depict scenes from the Old and New Testaments in 26 panels. They are dated to about 1360.

St Mary's Church, now used mainly as a concert hall, was built between 1348 and 1372. Originally Gothic, with aisles of the same height as the nave, it was remodelled in Baroque style in the 17th c. and subsequently restored to neo-Gothic. The concave steeples of 1967 (architect F. M. Černý) represent a compromise between Baroque and Gothic.

Trams
4, 6, 16, 18, 24

Metro
Karlovo náměstí

Bus route
148

Ethnographic Museum (Národopisné muzeum)　　　　　F2

The Ethnographic Museum, a branch of the National Museum, occupies the old Villa Kinsky, set in the Kinsky Gardens to the south of the Hunger Wall (see Petřín Hill). The collection includes models of peasant houses, reproductions of rooms in typical houses, pottery, costumes, embroidery and much else of interest.

Behind the museum are a picturesque little bell-tower of Moravian-Wallachian type and a small 18th c. Ukrainian wooden church.

Location
Petřínské sady 98, Smíchov, Praha 5

Opening times
Tues.–Sun. 10 a.m.–6 p.m.

---

*European Art Collection of the National Gallery　　D2
(Sbírska evropského umění/Národní galerie)

From Hradčany Square (see entry) a street runs west past the Archbishop's Palace to the Sternberg (Šternberk) Palace, now occupied by the National Gallery's Collection of European Art.

This High Baroque palace was designed by D. Martinelli and completed by J. B. Alliprandi in 1698–1707. It consists of four wings surrounding a courtyard, with a cylindrical projection. The walls of the courtyard have stucco ornament, and in the interior are ceiling-paintings by Pompeus Aldovrandini.

The collection includes works by painters of the Italian, Dutch and German schools (first and second floors) and by 19th and 20th c. painters from France and other European countries (upper floor and first floor). The rooms are not numbered but are arranged according to the artists' countries of origin.

Location
Hradčanské náměstí 15

Metro
Hradčanská

Trams
22, 23

Opening times
Tues.–Sun. 10 a.m.–6 p.m.

The first floor is mainly devoted to painters of the Florentine school of the 14th and 15th c. (Sebastiano del Piombo, Orcagna, Giovanni d'Allemagna, Antonio Vivarini, Piero della Francesca, Palma il Vecchio, etc.).

The narrow room opening off the third room (on the left) contains icons from different countries.

First floor

Dürer's "Festival of the Rosary"

Then follow Dutch and Flemish masters of the 15th and 16th c.: Geertgen tot Sint Jans, Gerard David, Jan Gossaert, known as Mabuse ("St Luke painting the Virgin"), Cornelis Engelbrechtsz, Pieter Brueghel the Elder ("Haymaking"), and Pieter Brueghel the Younger ("Winter Landscape", "Arrival of the Three Kings").

Second floor

The second floor begins with Italian painters of the 16th to 18th c., including works by Tintoretto ("David with Goliath's Head"), Veronese, Palma il Giovane, Bronzino, Tiepolo ("Portrait of a Venetian Patrician") and Canaletto ("View of London").

Then come German painters of the 14th to 18th c. The finest work here is Dürer's "Festival of the Rosary", painted in 1506 for German merchants in Venice. It shows the Virgin holding the Child and being crowned by angels, surrounded by numerous figures, including the artist himself (above, right), the humanist Konrad Peutinger, the wealthy German merchant Ulrich Fugger the Elder, the Emperor Maximilian I, Pope Julius II and a number of Venetians. There are also pictures by unknown 15th c. masters including the Master of Grossgmain and works by Hans of Tübingen, Hans Holbein the Elder, Hans Baldung Grien, Altdorfer, Lukas Cranach the Elder and Thomas Burgkmair.

There follow 17th c. Flemish and Dutch painters, including works by Jacob Jordaens ("Apostles"), Peter Paul Rubens ("Cleopatra", "Martyrdom of St Thomas" and "St Augustine on the Seashore", both 1637–39), David Teniers the Younger ("Boon Companions"), Antony van Dyck ("Abraham and Isaac"), Frans Hals ("Portrait of Jasper Schade van Westrum", 1645), Rembrandt ("Old Scholar", "Annunciation"), Adriaen van Ostade, Gerard Dou ("Young Woman on Balcony"), Metsu, Gerard Terborch (two portraits) and Philips Wouwerman.

The north and west wings are reached by keeping straight across the courtyard from the entrance.

North and West wings

French art of the 19th and 20th c., from the Romantic school to Cubism, arranged in chronological order according to the artists' dates of birth, with works by Delacroix, Daumier, Rousseau, Courbet, Monet, Cézanne, Renoir, Gauguin, van Gogh, Toulouse-Lautrec, Matisse, Vlaminck, Utrillo, Picasso (important paintings of the Early Cubist period), Braque and Chagall. Sculpture by Degas, Rodin, Bourdelle and Maillol.

Ground floor

On the first floor of the north and west wings is the collection of 19th and 20th c. European art (Sbírka evropského umění 19.a.20.století), with works by Austrian (Ferdinand Georg Wald-müller, Gustav Klimt, Oskar Kokoschka), German (Casper David Friedrich, "The Bridge"), Russian (Ilya Rjepin) and Italian (Renato Guttuso; sculpture by Manzù) artists.

First floor

Faust House

See Charles Square

Federal National Assembly   F5
(Areál budovy Federálního shromáždění ČSSR)

The modern building which houses the Federal National Assembly of the Czechoslovak Socialist Republic was erected in 1967–73 as a symbol of the planned development along the north–south axis road. It was designed by K. Prager, J. Albrecht and J. Kaderabek. Incorporated in the building is the former Stock Exchange (1936–38). Equally symbolic is the piece of sculpture in front of the building, Vincenc Makovský's "New Era". The interior is decorated with works by modern artists.
There are underpasses from the National Assembly to the Central Station and the Smetana Theatre (see Practical Information – Theatres).

Location
Vinohradská 1, Nové Město,
Praha 1

Metro
Muzeum

Tram
11

Franciscan Garden (Františkánská zahrada)

See St Mary of the Snows

Grand Prior's Palace (Palác maltézského velkopřevora)   E3

The old Grand Prior's Palace of the Knights of Malta in the Lesser Quarter (see entry) is now occupied by the Music Department of the National Museum (Hudební oddelení Národního muzea), with a fine collection of musical instruments and a musical archive (Muzeum české hudby; including pianos, organs, harps and other instruments). The impressive Baroque rooms in the museum have wooden panelling, ornate stoves from that period and splendid inlaid flooring.
During the summer months, concerts and theatre performances are frequently given in the garden.
The palace, with two wings set at right angles, was built in 1724–28 by the Italian architect Bartolomeo Scotti, who re-modelled an earlier Renaissance building and added a new door-way, ornamental cornices and oriel windows. The vases and

Location
Velkopřevorské nám. 4,
Malá Strana, Praha 1

Trams
12, 22

Opening times
Tues.–Sun. 10 a.m.–noon
and 1–5 p.m.

Grand Prior's Palace

Federal National Assembly of the ČSSR

Museum of musical instruments . . .   *. . . in the Grand Prior's Palace*

statues holding lamps featured in the stair-well are from the workshop of Matthias Braun.

Hibernian House (U hybernů) E5

The exhibition and market hall known as U hybernů was originally a Late Baroque church, built between 1652–59, belonging to a friary of Irish Franciscans who had resided here since 1629. The east and north wings were added in the 18th c.
After the dissolution of the friary (1786) and the church (1790), J. Zobel converted the latter into a customs post (architect, J. Fischer) with a neo-Classical façade (1808–11). The sculptural decoration was the work of F. X. Lederer (1811). At the beginning of the 1940s, J. K. Řiha re-fitted the interior of the building in order that it could be used for exhibitions and occasional cultural events.

Location
Náměstí republiky
(Republic Square),
corner of Hybernská, Staré
Město, Praha 1

Metro
Můstek

Trams
3, 5, 10, 26

High Synagogue

See Josefov

Holy Cross Chapel (Hradčany)

See Hradčany

Holy Cross Chapel (Old Town) (Rotunda svatého Kříže) E4

Situated just a short distance from the banks of the Vltava, the Chapel of the Holy Cross, built about 1100, is one of only three round chapels of the Romanesque period which survive in Prague. A proposal during the 19th c. to pull it down was frustrated by the opposition of the Czechs Artists' Union, and instead of being demolished it was renovated between 1863 and 1865 by the architect V. I. Ullmann and the painter Bedřich Wachsmann who also designed the new altar. The remains of Gothic wall-paintings in the interior of the chapel, representing the coronation of the Blessed Virgin, were restored by Sobislav Pinkas and František Sequenz. The iron grille was the work of Josef Mánes and the paintings on the triumphal arch and the apse are by Peter Maixner.

Location
Ulice Karoliny Světlé, corner
of Konviktská, Staré Město,
Praha 1

Trams
9, 17, 18, 21, 22

Holy Cross Church

See Na Příkopě

*House of Representation (Representačni dům hlavního města Prahy) E5

The Community House was built in 1906–11 (architects Antonín Balšanek and Osvald Polivka) on a site steeped in history. A royal palace was founded there in 1380 and later abandoned in 1547. It was there on 28 October 1918 that the Czech Republic was

Location
Praha 1
Staré Město
Náměstí Republiky

Representation House . . .   *. . . in imperial and royal style*

Metro
Müstek

Trams
5, 9, 19, 29

declared an independent state by the National Committee. It is today officially known as the House of Representation for the City of Prague. This fine building in the Art-Nouveau style with its traditional restaurants, cafés, wine-bars, exhibition halls, offices and the largest music hall in Prague (Smetana Hall) is a typical example of Czech Sezession buildings constructed at the turn of the century. In addition to the tendency towards rich decoration there was a special liking for geometric shapes and elaborate detail, plants and the ever recurring theme of youth.

A whole generation of artists was involved in the development of both the external and internal features of the building. The decoration on the façade and on the supports for the pillars on the balcony was by K. Novák. The ceiling sculpture in the Smetana Room was by the same artist, and L. Šaloun was responsible for the allegorical paintings depicting "Humiliation" and "Awakening of the Nation" on the front of the house. "Bohemian Dances" and the "Vysehrad" on the rostrum in the Smetana Room are further examples of his work. The symbolic fine art wall-paintings in the Primator's Room are by Alfons Mucha; pictures by M. Švabinský hang in the Rieger Room, and the Palacký Room contains a bust by J. V. Myslbek. Allegorical decoration may also be seen in the Grégor Room, while the Sladkovský Room has landscapes by J. Ullmann.

House signs

The dull and unimaginative method of identifying houses by giving them numbers, following the French model, was introduced in Prague as late as 1770, during the reign of Maria

House signs

Theresa. The use of house signs is much older. The use of family names (surnames) became general in the 13th and 14th c., and at the same time the practice was adopted of identifying houses by names which were represented pictorially on the outside of the building. At first the names usually referred to the situation of the house ("At the chestnut tree", "At the bridge") or the owner's occupation ("At the salt-house", "At the mill"), but came to refer also to features related to religion ("At the sign of the Black Virgin", "At the sign of the Golden Angel", "At the sign of Mary on the Golden Rock"), to plants ("At the sign of the three-leafed clover", "At the sign of the Black Rose", "At the sign of the Green Tree", "At the sign of the Golden Lily", "At the sign of the Golden Melon", "At the sign of the Three Red Roses") to animals ("At the sign of the Green Frog", "At the sign of the Black Pony", "At the sign of the Stone Lamb", "At the sign of the Little Bears", "At the sign of the Big Beehive", "At the sign of the White Unicorn", "At the sign of the Three Golden Lions", "At the sign of the Pelican") or to heavenly bodies ("At the sign of the Two Suns", "At the sign of the Blue Star"). The house retained its name even if there was a change of ownership.

The house signs which visitors will come upon as they walk about Prague may be of stone, metal or wood, or sometimes of stucco or painted on plaster or tin; occasionally they may have an inscription.

The signs were so popular that they continued to be used even after the introduction of street-numbering in 1770, during the Empire (for example at No. 11, the Melantrichova "At the sign of the Five Crowns") and neo-Classical periods (for example in the Lesser Quarter, "At the sign of the White Shirt"). Nowadays they are used mainly by restaurants and wine-houses.

There are so many houses with signs of this kind that they cannot possibly be listed here. They are mostly to be found in the back streets of the Old Town, but can also be seen in those recently restored (Celetná, e.g. No. 21, "At the sign of the Golden Vulture", No. 34 "At the sign of the Black Virgin"), in the Old Town Square (e.g. No. 24, Madonna with the Christ Child) or in the Karlova, e.g. No. 18, "At the sign of the Golden Snake"), as well as in the Lesser Quarter (e.g. "At the sign of the Three Ostriches"), where the picturesque Neruda Street leads to the Hradčany (Nerudova, e.g. No. 6 "At the sign of the Red Eagle", No. 12, "At the sign of the Three Fiddles", No. 16 "At the sign of the Golden Chalice", No. 34 "At the sign of the Golden Horseshoe", No. 36, "At the sign of Our Lady of Perpetual Succour", No. 49 "At the sign of the White Swan").

Hradčany (castle)   D2/3

Situation
North-west
Above the Vltava

Metro
Malostranská, Hradčanská

Tram
18, 22

Opening times
Apr.–Sept.,
Tues.–Sun. 9 a.m.–5 p.m.;
Oct.–Mar.,
Tues.–Sun. 9 a.m.–4 p.m.

Conducted tours

Hradčany Castle has been the official residence of the President of the Republic since 1918.

Hradčany was founded by the Přemyslids at the end of the 9th c. as a timber-built stronghold in three parts surrounded by a mud wall. In honour of St Vitus, Wenceslas the Holy had a Romanesque rotunda built between 926 and 929 on the site of the present Wenceslas Chapel. From 973 it was the residence not only of the ruling prince but of the bishop of the newly established diocese of Prague. In 1042, during the reign of Břetislav I, the castle was surrounded by a wall 2 m (6½ ft) thick and towers were built at the east and west ends; later a gateway was built on the south side. From 1135 onwards Soběslav I strengthened and developed the castle into a princely fortress in Romanesque style. The 30 m (100 ft) high Black Tower was used as a prison. During the reign of Ottokar II work was begun on the middle section of the former royal palace. In 1303 most of the structure was devastated by a fire. Work on the castle was resumed during the reign of Charles IV, in 1344. After the end of the Hussite Wars further alteration and rebuilding was carried out by the Jagellionian rulers and by Kings Vladislav (from 1471) and Louis (from 1516). During these building phases the first Renaissance features began to appear in combination with the Late Gothic features. The Emperors Ferdinand I (from 1556) who summoned artists to Prague from Italy, the Netherlands and Germany, and Rudolf II (from 1575) enriched the castle and surrounding area with magnificent Renaissance buildings. In 1614 the Emperor Matthias built Prague's first secular Baroque structure, the gate at the west end. Originally a free-standing tower, this was incorporated in the outer wall of the first courtyard in the 18th c. at the behest of Maria Theresa.

This building phase gave the Hradčany the architectural unity which makes it the dominant feature of the city's skyline to this day. After the declaration of the Republic in 1918 and the

View of the Lesser Quarter and Hradčany

Prague Castle

liberation of 1945 it was used for major State and public occasions and ceremonies.

In order to get an idea of the scale of the Hradčany and the multiplicity of buildings it contains, the best plan is to begin by walking round the various courtyards or outer wards with their little streets and lanes and then to visit the individual buildings.

Outer wards

First Courtyard
(První nádvoří)

The first and most recent of the courtyards, also known as the Grand Courtyard, is entered from Hradčany Square (Hradčanské náměstí; see entry) through a wrought-iron gateway flanked by piers bearing copies of Ignaz Platzer the Elder's "Fighting Giants" (1786).

The First Courtyard was created in the reign of Maria Theresa, between 1756 and 1774; the plans were prepared in Vienna by the Court Architect, Nikolaus Pacassi, and the work was directed by Anselmo Lurago. The sculptured trophies on the attics of the building are originals by Platzer. The most recent alterations to the courtyard were carried out by the Slovene architect Josip Plečnik in 1920–22.

Matthias Gate
(Matyasova brána)

The gate which bears the Emperor Matthias's name was built for him by Giovanni Maria Philippi in 1614 as a free-standing western entrance to the Hradčany. In 1760 the gate-tower was linked with the newly built front wall of the castle by N. Pacassi.

From the gate a flight of steps (by Pacassi, 1765–66) leads up to the State apartments of the castle – the Throne Room, a room containing paintings by Václav Brožik, the Hall of Mirrors, the Music Room and the Drawing Room. This range of buildings also contains the private apartments of the President of the Republic. The flagpoles outside the Matthias Gate are firs from the forests on the frontiers of Czechoslovakia – an idea conceived by the architect, J. Plečnik.

Second Courtyard
(Druhé nádvoří)

The Matthias Gate leads into the Second Courtyard, in the centre of which is a Baroque well-house, built by Francesco della Torre in 1686, with sculpture by Hieronymus Kohl. The wrought-iron grille dates from 1702.

The austerity of this courtyard is relieved by the modern Lion Fountain (V. Makovský, 1967) and the gleaming granite paving (J. Frágner, also 1967).

On the north side of the courtyard is the Plečnik Hall, created in 1927–31 by the reconstruction of older buildings, which was combined with the so-called Staircase Hall to form an entrance lobby to the Spanish Hall and the Rodulf Gallery.

From the Second Court a bridge (the Dusty Bridge; Prasný most) leads over the Deer-Pit; then, passing through St Mary's Work (Mariánské hradby) and continuing past the Royal Garden (open only in spring) and the old Ballroom, we come to Belvedere Palace (see entry).

Third Courtyard
(Tretí nádvoří)

The Third Courtyard was the centre of the castle's life. This was the starting-point of the main street of the old Slav settlement.

On the north side of the courtyard is St Vitus's Cathedral. Under the south wall of the cathedral can be seen the foundations (excavated 1920–28) of a Romanesque episcopal chapel. Between 1750 and 1770 the older buildings of the royal stronghold were given a uniform façade by N. Pacassi (Rudolf II's Renaissance palace, the Early Baroque Queen's Palace and the Palace of Maximilian II). Under the balcony with statues holding lamps (by

Hradčany: Matthias Gate

Golden Lane

Ignaz Platzer) is the entrance to the offices of the President of the Republic.

The Golden Lane (also called the Gold-Makers' Lane) runs between the castle walls built by Vladislav Jagiello and the Old Castellan's Lodging. Originally it continued to St George's Convent. Along its north side ran the wall-walk between the Daliborka Tower and the White Tower. The north side of the street has been preserved, with its tiny picturesque houses built into the arches under the wall-walk. Rudolf II assigned these houses to the 24 members of his castle guard, who followed various trades in their leisure time. The name of Gold-Makers' Street or Alchemists' Street refers to Rudolf II's alchemists, who are traditionally said to have lived and sought to produce gold in these houses. It is known, however, that their laboratories were in the Mihulkar Tower. The houses were later occupied by craftsmen and the poorer members of the community. In 1912–14 Franz Kafka lived and wrote in house No. 22. The houses are now occupied by small shops (souvenirs, etc.).

Golden Lane
(Zlatá ulička)

The former Castellan's Lodging is now the House of Czechoslovak Children. Exhibitions, recitals and concerts take place in the recently renovated Lobkovitz Palace.

South-east end of castle precincts

To the west of the Bastion is the Rampart Garden (views). Two obelisks mark the spots where the Emperor's representatives fell when thrown out of the window in the Second Defenstration of Prague (1618). Above the New Castle Steps is the Paradise Garden, with the Matthias Pavilion and a music pavilion.

*Castle Gallery (Hradní galerie)

The Castle Gallery was created in 1965 by converting the old

71

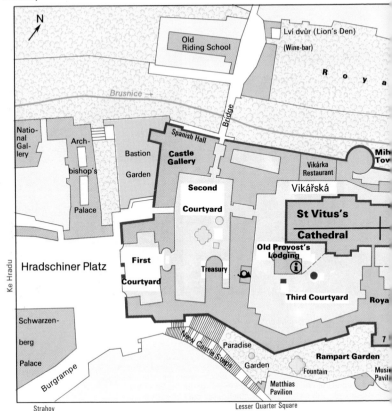

court stables in the north wing and the ground floor of the west wing. Its six rooms contain a total of 70 pictures from the old Rudolf Gallery and the Castle Gallery established by Ferdinand II and later broken up. Among the most notable works are a portrait of the Emperor Matthias painted by Hans of Aachen about 1612, Titian's "Young Woman at her Toilet", Tintoretto's "Woman taken in Adultery", Veronese's "St Catherine with Angel" and Rubens's "Assembly of the Olympian Gods" (c. 1602). The collection also includes pictures by Bohemian artists of the Baroque period (Jan Kupecký. Johann Peter Brand) and sculpture by Adriaen de Vries ("Adoration of the Kings") and Matthias Braun. The remains of a 9th c. church, the first to be built in Hradčany were discovered in the area of the picture gallery by the archaeologist I. Borkovský.

Chapel of the Holy Cross (Kaple svatého Kříže)

At the south corner of the Second Courtyard is the Chapel of the Holy Cross, which since 1961 has housed the Treasury of St

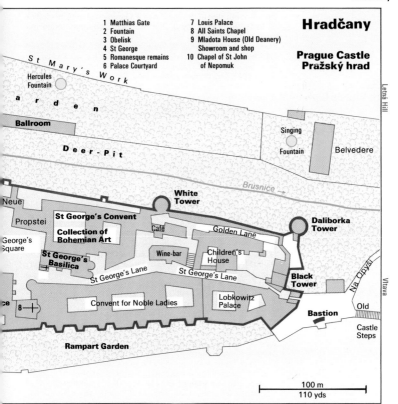

1 Matthias Gate
2 Fountain
3 Obelisk
4 St George
5 Romanesque remains
6 Palace Courtyard
7 Louis Palace
8 All Saints Chapel
9 Mladota House (Old Deanery) Showroom and shop
10 Chapel of St John of Nepomuk

Vitus's Cathedral. It contains valuable liturgical utensils, vestments, monstrances and reliquaries, St Wenceslas's coat of mail, St Stephen's sword and other relics.

The chapel was built in 1756–63 by Anselmo Lurago, but was altered between 1852 and 1858, during the Biedermeier period, with the idea of relieving its neo-Classical severity. The statue of St John of Nepomuk in the interior and those of SS Peter and Paul in niches are by E. Max (1854). The wall and ceiling paintings by J. Navrtil and V. Kandler are from the same period. The windows are the work of J. Z. Quast and the sculpture of the high and side altars is by Ignaz Platzer. The painting of the Crucified Christ in the centre of the high altar is by F. X. Balko.

Old Provost's Lodging (Staré proboštsví)

At the south-west corner of St Vitus's Cathedral is the Old Provost's Lodging. Originally a Romanesque episcopal palace, it was given its present Baroque form in the 17th c. The statue of St Wenceslas is by Johann Georg Bendl (1662).

On the south side of the Old Provost's Lodging stands an obelisk of Mrákotín granite (by J. Plečnik, 1928) commemorating the dead of the First World War.

The equestrian statue of St George (a copy: the original is in St George's Convent), in Early Gothic style, was the work of Georg and Martin of Cluj (1373); it was restored by Tomáš Jaros after a fire in 1541. The present base is by J. Plečnik (1928).

**St Vitus's Cathedral (Chrám svatého Víta)

St Vitus's Cathedral, the metropolitan church of the archdiocese of Prague, stands on the site of a round chapel which Duke (St) Wenceslas dedicated to St Vitus in 925.

One hundred and thirty-five years later Duke Spytihněv II founded a Romanesque basilican church with a double choir. In 1344 Charles IV began the construction of the present Gothic cathedral.

The east end was designed by the French architect Matthias of Arras, following the older French Gothic style (Narbonne and Toulouse Cathedrals). He was responsible for the choir (47 m (154 ft) long, 39 m (128 ft) high), though only the lower part of it was completed when he died in 1352. His successor, Peter Parler, enriched the structure with the upward-soaring German Gothic forms. The work was continued by his sons Wenzel and Johann (1399–1420), who completed the choir with its ring of chapels and built the lower part of the main tower. After the Hussite Wars Bonifaz Wohlmut and Hans of Tirol topped the tower with a Renaissance steeple and balustrade (1560–1562), bringing it to a total height of 109 m (358 ft); and finally, in 1770, it was given its onion dome by N. Pacassi.

St Vitus's Cathedral was only finished in the early 20th c. Following the design by Peter Parler, Josef Mocker worked on the cathedral from 1872 onwards, beginning with the neo-Gothic western part including the main portal. This was completed in 1929 by Kamil Hilbert.

St Vitus's Cathedral is not only Prague's most imposing church and the finest building in the Hradčany: it is also the city's largest church, with a total length of 124 m (407 ft), a breadth of 60 m (197 ft) across the transepts and a height of 33 m (108 ft) in the nave. In the south tower – an unusual combination of Gothic, Baroque and Renaissance elements – are three Renaissance bells and the largest church bell in Bohemia, the bronze Sigismund Bell (1549).

Interior

Visitors enter the cathedral by the west doorway. The south doorway is closed at present owing to renovation work. On the upper part of this sumptuous portal, known as the Golden Gate (Zlatá brána), are portraits of Charles IV and Elisabeth of Pomerania and a much-restored 14th c. mosaic of the Last Judgment. Above this is a traceried window by Max Svabinský (1934), also representing the Last Judgment, which contains no fewer than 40,000 separate pieces of glass.

Triforium gallery

The best plan is to leave the side chapels for the moment and get a general impression of this massive building with its 28 piers and 21 chapels, from the centre of the cathedral.

The triforium gallery runs above the arcading and below the windows of the choir. Within the triforium, and particularly over the organ gallery, are busts of the cathedral architects, Charles IV's family and other notable personages. These busts, mostly from the Parlers' workshop, formed the first gallery of portraits of

St Vitus's Cathedral and Old Provost's lodgings ▶

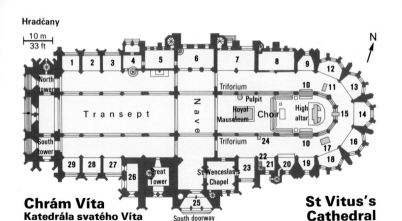

Chrám Víta
Katedrála svatého Víta

South doorway

St Vitus's Cathedral

1 Bartoň-Dobenín Chapel
2 Schwarzenberg Chapel
3 New Archbishops' Chapel (Hora Chapel)
4 Old Treasury
5 New Sacristy
6 Wohlmut's choir (organ gallery)
7 St Sigismund's Chapel (Czernin Chapel)
8 Old Sacristy (formerly St Michael's Chapel)
9 St Anne's Chapel (Nostitz Chapel)
10 Historical reliefs

11 Statue of Cardinal von Schwarzenberg
12 Old Archbishops' Chapel
13 Chapel of John the Baptist (Pernstein Chapel)
14 Lady Chapel (Trinity Chapel, Imperial Chapel)
15 Tomb of St Vitus
16 Reliquary Chapel (Saxon Chapel, Sternberg Chapel)
17 Tomb of St John of Nepomuk
18 Chapel of St John of Nepomuk (St Adalbert's Chapel)

19 Waldstein Chapel (Magdalene Chapel)
20 Vladislav Oratory (Royal Oratory)
21 Holy Cross Chapel
22 Entrance to Royal Vault
23 St Andrew's Chapel (Martinitz Chapel)
24 Monument of Count Schlick
25 Crown Rooms
26 Chapter Library
27 Thun Chapel
28 Chapel of Holy Sepulchre
29 St Ludmilla's Chapel (Baptistery)

historical figures in Europe before the Renaissance. There is no access to the triforium, but there are casts of the busts in the Royal Palace (see p. 58) and in Karlštejn Castle (see entry).

Organ gallery
The two-storey organ gallery (by Bonifaz Wohlmut, 1557–61) is opposite the south doorway. After the completion of the cathedral it was moved from its original position at the west end to the north aisle. The organ (1757) has 6500 pipes.

Imperial Mausoleum
In the centre of the choir, in front of the high altar, is the white marble Imperial Mausoleum, surrounded by a Renaissance screen (by J. Schmidthammer, 1589). The mausoleum was begun in Innsbruck in 1564 as a memorial to Ferdinand I and his wife Anna Jagiello and remodelled in the reign of Rudolf II (completed 1589). The figures on the cover are Anna Jagiello, Ferdinand I (centre) and Maximilian II. In medallions along the sides are representations of the kings and queens of Bohemia who are buried in the vault under the mausoleum.

Burial vault
The entrance to the vault is beside the Holy Cross Chapel. In the passages archaeological finds of the Romanesque period are displayed. On the wall is a ground-plan of the old Romanesque church.
The sarcophagi in the upper tier are those of George of Poděbrad (1420–71: left), Charles IV (1316–78: centre) and Ladislav Postumus (1440–57: right). In the lower tier are Wenceslas IV (1361–

1419), his brother John of Görlitz (d. 1396) and the common sarcophagus of Charles IV's four wives. To the rear is Maria Amalia, Maria Theresa's daughter. A Renaissance sarcophagus in pewter contains the remains of Rudolf II (1552–1612), and a low granite sarcophagus those of Charles IV's children. The royal crypt was reorganized between 1928 and 1935.

Old Sacristy
Visitors leave the burial vault by a staircase which emerges in front of the choir screen. To the left is the Old Sacristy, which formerly contained the cathedral treasury (now in the Holy Cross Chapel in the Second Courtyard). Note the fine stellar vaulting.

St Anne's Chapel
Immediately behind the Old Sacristy is St Anne's Chapel, opposite which is the first part of a relief wood-carving attributed to Caspar Bechterle of Niedersonthofen depicting the destruction of images in the cathedral in 1619; its counterpart on the south side of the ambulatory depicts the flight of the "Winter King", Frederick V of the Palatinate, after the Battle of the White Mountain. Also of interest is a view of Prague as it was at that period (drawn about 1630).
Just beyond St Anne's Chapel is a kneeling figure, in bronze, of Cardinal Friedrich von Schwarzenberg (d. 1885) by Josef Václav Myslbek (1895).

Archbishops' Chapel
Opposite the Schwarzenberg statue is the Archbishops' Chapel, with the burial vault of the archbishops of Prague.

Chapel of John the Baptist
This adjoins the Archbishops' Chapel. To the right and left are the

Tomb of St John of Nepomuk   *Bishop's throne*

tombs of Břetislav II (d. 1100) and Bořivoj II (d. 1124). The bronze candelabra to the left of the altar is said to have come from Vladislav II's share of the booty brought back by Frederick Barbarossa from Milan. The Romanesque base of the candelabra came from the Rhineland.

Lady Chapel
In the Lady Chapel are the tombs of Břetislav I (1034–55) and Spytihněv II (1055–61), both from the Parlers' workshop. Opposite the chapel, behind the high altar, can be seen the Tomb of St Vitus, with a statue of the Saint by Josef Max.

Reliquary Chapel
This chapel, built by Charles IV and also known as the Saxon Chapel, contains the tombs of Přemysl Ottokar I (1192–93 and 1197–1230), on the right, and Přemysl Ottokar II (1253–78), on the left. Both are from the Parlers' workshop.
Chapel of St John of Nepomuk, next to the Reliquary Chapel; also known as St Adalbert's Chapel. On the altar are silver busts of SS Adalbert, Wenceslas, Vitus and Cyril (c. 1699).

Opposite the chapel is the sumptuous silver Tomb of St John of Nepomuk, made in Vienna in 1733–36 by A. Corradini and J. J. Würth to the design of Joseph Emanuel Fischer von Erlach.

Opposite the next chapel, the Waldstein or Magdalene Chapel (with the Waldstein family vault), is the second part of the relief by Caspar Bechterle (above, p. 77).

Royal or Vladislav Oratory
This chapel is now attributed to a Frankfurt sculptor named Hans Spiess. It is a richly decorated Late Gothic structure with a

Tracery by Švabinský

Art-Nouveau window

naturalistic pattern of interwoven branches on the front, formed of two intersecting arches with a pendant boss.

Chapel of the Holy Cross
On the left-hand wall of this chapel is a painting of the Vernicle (1369), with representations of the six patron saints of Bohemia on the frame. Entrance to the Royal Burial Vault.

St Andrew's Chapel
In this chapel, also known as the Martinitz Chapel, is the grave-stone (to the left, under the window) of Jaroslav von Martinitz (d. 1649), one of the two victims of the Second Defenestration of Prague.

Monument of Count Schlick
On the first pier opposite St Wenceslas's Chapel is the Baroque monument of Field-Marshal Count Schlick (d. 1723), by Matthias Braun (to the design of J. E. Fischer von Erlach).

St Wenceslas's Chapel
The finest of the choir chapels is the Gothic Chapel of St Wences-las, which extends into the south transept. It was built by Peter Parler in 1358–67, replacing the original round chapel of the Romanesque period in which the Saint was buried. It contains the Shrine of St Wencelsas, Duke of Bohemia, who was murdered by his brother Boleslav in 929 or 935. The lion's-head door-ring to which, according to the legend, he clung when attacked by Boles-lav is preserved here.
The lower part of the walls of the chapel is decorated with 1300 Bohemian semi-precious stones. The lower register of wall-paintings (the Passion cycle) is by Master Oswald of Prague (1373); the upper register (legend of St Wencelsas) is from the studio of the Master of the Litoměřice Altar (c. 1509). The poly-chrome painting of the statue of St Wenceslas on the east wall between two angels (Heinrich Parler, 1373) is by Master Oswald. The bronze candelabra at the left-hand end of the wall was the work of Hans Vischer of Nuremberg (1532).

Crown Rooms
From St Wenceslas's Chapel a staircase leads up to the Crown Rooms over the south doorway in which the Bohemian Crown Jewels and insignia are kept. They are open to the public only at certain specified times.

**Royal Palace (Královský palác)

The Royal Palace, the history of which down the centuries is reflected in its architecture, is in the Third Courtyard. It was the royal residence until the end of the 16th c., but when, under the Habsburgs, the royal residence moved west it was used as offices and store-rooms.
Part of the original Romanesque palace has survived in the ground and basement floors below the present Vladislav Hall. The remains of the original palace were concealed under new building by Přemysl Ottokar II, Charles IV and Wenceslas IV, and later Vladislav Jagiello built or rebuilt a new floor, in the centre of which is the Vladislav Hall, the most splendid of the secular buildings in the Hradčany.

Green Room
The central doorway under the balcony, to the right of the Eagle Fountain (by Francesco Torre, 1664), leads into an antechamber,

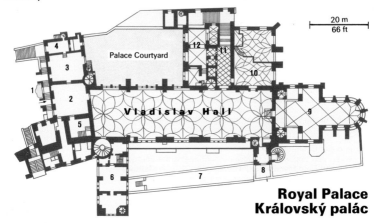

1 Eagle Fountain
2 Antechamber
3 Green Room
4 Vladislav's Bedroom

5 Romanesque tower
6 Bohemian Chancellery
7 Theresian range
8 Outlook terrace

9 All Saints Chapel
10 Hall of Diet
11 Staircase (for horsemen)
12 New Appeal Court

**Royal Palace
Královský palác**

with the Green Room to the left. From the time of Charles IV this was used as a court room and from the 16th c. was the seat of the Supreme Court and the Palace Court. On the east side, the walls are decorated with coats-of-arms of Upper and Lower Lusatia, and in addition there are several coats-of-arms which serve as a reminder of the old 18th c. assessors. The Baroque ceiling-fresco of the Judgment of Solomon (copy) was installed here in 1963. It was originally in the Castellan's Lodging which is now the House of Czechoslovak Children. The Green Room gives access to Vladislav's Bedroom and the Map Repository. Not only the coats-of-arms of Bohemia, Moravia, Silesia and Luxembourg may be seen in the bedroom but above the window there is also the royal monogram of Vladislav Jagiello. In the Repository are the coats-of-arms of the Lord High Chamberlain.

Vladislav Hall (Vladislavský sál)
The Vladislav Hall, also known as the Hall of Homage, was built by Benedikt Ried between 1493 and 1503. With its considerable dimensions (62 m (203 ft) long, 16 m (52 ft) wide, 13 m (43 ft) high) and its beautiful Late Gothic reticulated vaulting it is one of the show-pieces of the Hradčany. In this hall the Bohemian kings were elected, the provincial Diet held its meetings and jousting tournaments took place. Since 1934 it has been used for the election of the President of the Republic and in 1960 the Socialist basic law of the ČSSR was passed here by the National Assembly. A staircase leads to a gallery from which the interior of All Saints' Chapel may be viewed.

Hall of the Diet
The doorway at the north-east corner of the Vladislav Hall leads into the Hall of the Diet, and Common Law Chamber, also built by Benedikt Ried (c. 1500). After being devastated by fire it was rebuilt in 1559–63 by Bonifaz Wohlmut with ribbed vaulting in imitation of Late Gothic architecture.

Vladislav Hall

All Saints Chapel

Hall of the Diet

There are busts of the builder and his client, Emperor Ferdinand I and in the north-west corner the rostrum for the Clerk to the Diet is Renaissance. On the walls are portraits of Habsburg rulers. The tiled stove (neo-Gothic) near the entrance was made in 1836. The neo-Gothic throne (19th c.) stands between the windows; the heraldic lions above it are from the 17th c. Religious dignitaries and senior civic officials once sat to the right of the throne whilst noblemen and knights sat opposite. The balustrade to the right near the window was reserved for representatives of the royal towns. Prior to 1847, sessions of the Supreme Regional Court and of the Estates Diet were held here. Nowadays, after his election, the President of the ČSSR signs his pledge in this room and on ceremonial occasions the Czech Representative Assembly meets here.

New Map Room
The entrance to this room, which dates in its present form from the 17th c., is to the left of the doorway into the Hall of the Diet. On the walls are the coats-of-arms of officials of the Map Repository.

Staircase
This is no ordinary staircase but one with Gothic ribbed vaulting specially designed for horsemen, who could thus ride up into the Vladislav Hall for the tournaments which were held there.

Palace Courtyard
A doorway on the right at the foot of the staircase leads into St George's Square, the one on the left into the Palace Courtyard, surrounded on two sides by arcades. Another staircase gives access to the Gothic Palace (not at present open to the public).

Other features of the palace
Old Map Room, with massive vaulting borne on two squat piers; an arcaded passage from the time of Přemysl Ottokar II; the Charles Room, with casts of the Parler busts in the triforium of St Vitus's Cathedral; the Old Registry (or Palace Kitchen); and Wenceslas IV's Columned Hall, with late 15th c. Gothic vaulting.

Basement
A steep staircase leads down from the Palace Courtyard to the basement of the palace, in which visitors can see remains of early castle walls, some dating from the 9th c. (The basement and the adjoining Soběslav Palace, of the Romanesque period, are not at present open to the public.)

Louis
Palace

The Louis Palace which adjoins the Vladislav Hall was built by Benedikt Ried between 1502 and 1509 for Vladislav II. It contains the apartments occupied by the Bohemian Chancellery. The larger room with Gothic vaulting was the seat of the Governor of Bohemia. The smaller Council Chamber (1509), which bears Louis' monogram, is entered from this room through a Renaissance doorway.

From the windows of the room in the tower, on 23 May 1618, the Emperor's Deputy Governors Jaroslav von Martinitz and Wilhelm Slavata, together with their clerk Fabricius, were thrown 15 m (50 ft) down into the castle moat (but, according to a contemporary account, "got off with the fear of death, their lives and a few scratches"). This Second Defenestration of Prague gave the signal for the Bohemian rising against the Habsburgs and led to the Thirty Years' War.

From here a spiral staircase leads to the former Chancellery of the Imperial Council. It was here on 19 June 1621 that those who took part in the uprising by the Estates against the House of Habsburg received their death sentences. The entrance has inlaid work from the 17th c. and the Chancellery is decorated in Late Renaissance style. Above the door there is a picture by I. Raab depicting the siege of Prague by the Prussians in 1757. To the left, the portrait of the Spanish king Philip IV is the copy of a painting by Velázquez. The two wooden ceremonial shields above the stove are from the 17th c. and portray the funerals of Maximilian II and Rudolf II.

Another spiral staircase (not at present open) leads back to the Vladislav Hall.

All Saints Chapel
A short flight of steps at the east end of the Vladislav Hall gives access to the All Saints Chapel (by Peter Parler, 1370–87). Following a fire in 1541, Queen Elisabeth, daughter of Maximilian II, had the chapel rebuilt and extended (1579–1580) to join up with the Vladislav Hall. The chapel is entered through a Renaissance doorway. The high altar (c. 1750) built by P. Prachner has an All Saints altar-piece by W. L. Reiner. The Descent from the Cross on the right side altar is a Late Renaissance work of the Rudolf school. Hans of Aachen's Angels triptych beneath the gallery was created in the 16th c. The carved tomb (1739) with the remains of St Prokop is the work of F. I. Weiss. A cycle of paintings by Christain Dittmann on the lives of the saints (1669) decorates the walls of the choir.

The chapel originally belonged to the convent for noble ladies which adjoined the palace on the east.

St George's Basilica (Bazilika svatého Jiří)

At the east end of St George's Square, facing the chancel of St Vitus's Cathedral, is the twin-towered Romanesque Basilica of St George, the oldest surviving church in the Hradčany, which is now used as a concert hall.

The church was founded by Duke Vratislav I in 912 and consecrated about 925. It was devastated by fire in 1142 and again in 1541, and was much altered and rebuilt in the course of its history. The present Baroque façade dates from about 1670.

During renovation work in 1897–1907 and 1959–62 the original Romanesque character of the church was restored: the exterior marked by its slender white towers, the interior by the alternation of pillars and columns and the triple-arched galleries in the thick walls over the arcading.

South doorway
The south doorway (c. 1500), in Early Renaissance style, came from the workshop of Benedikt Ried; it is decorated with a copy of a Late Gothic relief of St George.

Nave, choir and crypt
The nave has 12th c. gallery windows and its original arches which date back to the 10th/11th c. The raised choir has the remains of Romanesque ceiling paintings ("Heavenly Jerusalem", after 1200). Late Renaissance frescoes (16th c.) on the ceiling of the apse depict the Coronation of Our Lady. In front of the entrance to the crypt is the monument of Duke Boleslav II (d. 999), enclosed within wrought-iron Baroque screens (c. 1730). To the right is the painted wooden tomb of Vratislav I (d. 921). The crypt was constructed in the middle of the 12th c. and contains a statue of St Brigitte by B. Spinetti.

Façade of St George's Basilica

Chapel of St Ludmilla
Built in the 13th c., the Chapel of St Ludmilla was altered in the 14th c. to house the reliquaries. The Renaissance vault contains the tomb made by Peter Parler about 1380 of the country's patron saint, Ludmilla (mudered in 921). Her life is portrayed in a fresco (1858) by J. V. Hellich on the west wall of the chapel. Other paintings from the end of the 16th c. depict Christ, the Blessed Virgin Mary, the Evangelists and Bohemian sovereigns.

Chapel of St John of Nepomuk
Built on to the south wall of the church is the Chapel of St John of Nepomuk (by F. M. Kaňka, 1718–22). The statue of the Saint on the façade is by Ferdinand Maximilian Brokoff. Notable features of the interior are the frescoes in the dome – Apotheosis of the Saints – and the altar-piece by V. V. Reiner. The statues of St Adalbert and St Norbert which stand in the niches in the Chapel date back to about 1730.

*Collection of Old Bohemian Art of the National Gallery
(Sbírka starého českého umění/Národní galerie)

Opening times
April–Nov.:
Tues.–Sun.
10 a.m.–6 p.m.

Adjoining St George's Basilica stands the Benedictine Convent of St George, a nunnery founded in 973 by Duke Boleslav II and his sister Mlada, who was the convent's first abbess. This pre-Romanesque (Ottonian) building is the oldest religious house in Bohemia. After being damaged by fire in 1142 and again in 1541 it was much altered, enlarged and remodelled in Baroque style (1657–80). It was dissolved in 1782, and now houses the National Gallery's collection of old Bohemian art.

Lower floor

The collection is arranged chronologically, beginning with Gothic on the lower floor. The sculpture and panel-paintings are

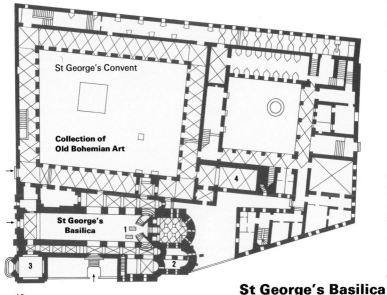

St George's Convent

Collection of
Old Bohemian Art

St George's
Basilica

3

2

| 10 m |
| 33 ft |

1 Tombs of Přemyslid rulers   3 Chapel of St John of Nepomuk
2 St Ludmilla's Chapel   4 St Anne's Chapel

St George's Basilica and Covent
Bazilika a klášter svatého Jiří

mostly from churches in Bohemia. The artists, whose names are unknown, are identified by reference to their works and to the places where they were found. The commonest theme in medieval Bohemian art is the Virgin.

Thanks to the munificent patronage of Charles IV Gothic art flourished in Bohemia, and continued to thrive after his reign – for example in the Prague school of painters, represented here by their finest works.

North corridor
The long north corridor is dominated by the monumental tympanum from the Church of St Mary of the Snows (see entry; 1346) and Early Gothic Madonnas.

Cycle of the Master of Hohenfurth (Vyši Brod)
A separate room is devoted to the Cycle of the Master of Hohenfurth (Mistr vyšebrodského oltáře), an altar-piece of nine panels (*c.* 1330–50) from the Cistercian house of Hohenfurth (now Vyši Brod): Annunciation, Nativity, Crucifixion, Mourning, Resurrection, Ascension into Heaven, Christ on the Mount of Olives, Adoration of the Kings, Effusion of the Holy Ghost.

Equestrian statue of St George
Farther along the north corridor is the original of the statue of St George in the Third Courtyard. This bronze figure, cast by Martin and George of Cluj (Klausenburg) in 1373, is the earliest freestanding piece of sculpture north of the Alps.

Room of Master Theoderich

This room adjoins the north corridor. Theoderich, the only painter represented here whose name is known, is a representative of the style known to German art scholars as the "beautiful" or "soft" style of Gothic painting in Bohemia. He painted 128 panel-paintings for Charles IV in Karlštejn Castle (see entry), five of which can be seen here (St Elisabeth, St Vitus, St Hieronymous, St Matthew, Pope Gregory; mid 14th c.). The votive picture of Archbishop John Očko of Vlašim (c. 1370) with portrait-like representations of Charles IV and Wenceslas IV and the "Crucifixion" from Emmaus Abbey (see entry) reach out beyond the "soft" style.

Corridor

The long corridor displays fine statues of the late 14th c. Smaller items are shown in cases.

Ground floor

Cycle of the Master of Wittingau

The first room on the ground floor is devoted to the Cycle of the Master of Wittingau (Mistr třeboňského oltáře), of which there are only fragmentary remains. The three panels on the winged altar (only open on public holidays) depict: on the front Christ on the Mount of Olives, the Resurrection and the Entombment; on the back a cycle of saints (St Catherine, St Mary Magdalen, St Margaret, St Giles, St Gregory, St Hieronymous; the Apostles, James the Younger, Bartholomew and Philip), also from the workshop of the Master of Wittingau (who himself painted only the heads).

Tympanum from the Týn Church

A room opening off the north corridor displays the tympanum from the north doorway of the Týn Church (see entry; Peter Parler's workshop, 1402–10), with three scenes from the Passion cycle – the Scourging, Calvary and the Crowning with Thorns.

North corridor

The north corridor contains paintings and sculpture illustrating the development of Late Gothic art in Bohemia, with special reference to the "soft" style. Notable among the representations of the Virgin are the Madonna suckling the Infant Christ from Konopiště (c. 1380), the Madonna of Český Krulov (c. 1400) and the Madonna of Zlatá Koruna (1410).
The Crucifixion by the Master of Raigern (early 15th c.), with its almost caricatural distortion of the figures, marks a turning-point in Bohemian painting.

Renaissance works

Notable among the Renaissance works in the collection are the "Visitation" (1505) by the Master of Leitmeritz (Litoměřice), the "Lamentation" from Zebrák and the carved wood altar shrine by Master IP (known only by his initials; c. 1520) one of the unknown artists influenced by Albrecht Dürer, which is in the last room before the stairs up to the first floor.

First floor

The works on the first floor show the development from Mannerism (end of 16th c.) through Baroque and Rococo to the end of the 18th c.

Mannerism

At the beginning of this section are the artists belonging to the northern school of Mannerism at the Court of Rudolf II (1576–1611): the Dutch painter, Bartholomäus Spranger (1546–1611;

"The Resurrection of Christ Carrying His Cross", 1576; "Epitaph of Nikolaus Müller the Goldsmith", c. 1590), Hans von Aachen, painter to the Prague Court (1552–1615; "Christ Carrying His Cross", 1587), the Swiss painter Josef Heintz Roelandt Savery (1564–1609; "The Last Judgment", 1607) and the Dutch sculptor Adriaen de Vries (1560–1627; bronze figure of Hercules).

Baroque
The most prominent works by the Baroque painters are: Karel Škréta (1610–1674; "Vestalin Tuscia", c. 1603; "The Baptism of Christ" c. 1660), Johann Peter Brandl (1669–1735; "Healing of Tobias, the Blindman" c. 1795; "Portrait of an old man", c. 1720) and by the portrait painters Johannes Kupetzky (1667–1740; "Portrait of his wife", 1711; "Portrait of a young girl of Prague", 1716) as well as Wenzel Lorenz Reiner (1689–1743; "Turkish Battle", 1708; "Landscape with waterfall and fishermen", c. 1740), in addition the works of Michael Leopold Willmann (1630–1706; "The release of the Andromeda", c. 1695; "Crucifixion", c. 1700) who was artist to the Prussian Court.
Important works by sculptors include Ignaz Franz Platzer (1717–1787; "St Barbara", c. 1750), Matthias Bernhard Braun, Born in the Tirol (1684–1738; old Athenian statues in the Palais Clam-Gallas (see entry): "Jupiter", "Vulcan", "Venus", "Mercury", 1714–1716): Ferdinand Maximilian Brokoff (1688–1731; "The Archangel Raphael", c. 1724) and Johann Georg Bendl (1630–1680; "The Archangel Raphael", c. 1650).

Rococo
The development from Baroque to Rococo is seen in pictures by Anton Kern (1709–1747; "St John on Patmos", "The Upbringing of the Virgin Mary") and the genre artist Norbert Grund (1717–1767; "The Solemn Girl" and "The Artist's Studio").

---

*Hradčany Square (Hradčanské náměstí)   D2

The little town of Hradčany was founded about 1320, the third settlement on the site of Prague. It did not enjoy independent municipal status, subject only to the Crown, but owed allegiance to the Castellan of the castle on Hradčany Hill. Originally it covered only the area of the present Hradčany Square, but in the reign of Charles IV it was extended towards the north and surrounded by walls.

Hradčany Square, with its Baroque Plague Column (Statue of Mary with the eight patron saints) by Ferdinand Maximilian Brokoff (1725) and an interesting cast-iron candelabrum from the age of gas lighting (19th c.) forms the approach to Prague Castle (see entry) and was once the centre of the town of the Hradčany. It lay on the processional route, which began in the Vyšehrad (see entry), which was followed at the coronation of the kings of Bohemia. Here in 1547 the leaders of the abortive rising against the Habsburg King Ferdinand I were executed. Although it has the scale and layout of a medieval market-place it never performed this function.

After the city fire in 1541 the square was completely rebuilt, all the old burghers' houses being pulled down to make way for palaces.

In 1372, Peter Parler, the famous master-builder who was responsible for St Vitus's Cathedral, lived at No. 10. The Hržan Palace in Loreto Street also once belonged to him. In the 18th c. the Saxon

Location
Hradčany, Praha 1

Metro
Hradčanská

Trams
22, 23

Parler House

Candelabra in Art-nouveau style

Hržan Palace

residence together with the adjoining property, No. 9, owned at one time by the Rosenberg family, was given a completely new façade.

Palais Toscana (Toskánský palác)

This two-storey palace with four wings (No. 5) was built by the French architect, Jean B. Mathey between 1689 and 1691 for Michael Oswald Thun-Hohenstein. From 1718 to 1918 it belonged to the Dukes of Tuscany and is now the property of the Czech Foreign Ministry. The harmony of its proportions is enhanced by the way in which it is distanced from its surroundings. The Early Baroque façade has two doorways flanked by columns, the coats-of-arms of the Dukes of Tuscany above the balconies and six Baroque statues, in liberal art form by Johann Brokoff, on the attic storey. The corner statue of St Michael locked in battle with the dragon is the work of Ottavio Mosto (1693).

Archbishop's Palace (Arcibiskupský palác)

Opening times
Only on Maundy Thursday
9 a.m.–5 p.m.

The recently restored Archbishop's Palace on the north side of the square was originally a Renaissance mansion which Ferdinand I bought from the royal Private Secretary Florian von Gryspek and presented to the first post-Hussite Catholic Archbishop of Prague. It was rebuilt in 1562–64 to the design of H. Tirol. About 1600 it was extended and in 1675–1688 remodelled in Baroque style, with a magnificent new doorway, by the French architect Jean-Baptiste Mathey. It was given its present Rococo form, with marble facing on the façade, by Johann Joseph Wirch (1763–64), who was also responsible for the sumptuous interior decoration. The family crest of Archbishop Anton Peter Graf, Prince of Přichowitz may be seen above the pieces by the sculptor I. F. Platzer, of which "Faith" and "Hope" were replaced in 1888 by new works by T. Seidan.

Nine Gobelins produced in the Paris workshop of Neilson by A. Desportes have as their theme "India – old and new". Special mention however must be made of the ornate wood-carvings, stucco decoration, two Gothic reliquary busts of St Peter and St Paul in the chapel as well as the collection of glass and porcelain.

Martinitz Palace

At the north-west corner of the square is the Martinitz Palace, which is the residence of the capital's Chief Architect, where exhibitions, concerts and literary recitals are held. This handsome Renaissance palace was originally built for Andreas Teyfl at the end of the 16th c. as a plain building of four wings set round a courtyard. In 1624 it was acquired by Jaroslav Borita von Martinitz, known to history as one of the two Imperial councillors who suffered defenestration in 1618. He added an extra storey, together with the Renaissance gables and coats-of-arms. The east front has figural sgraffito decoration depicting scenes from the life of Samson (after 15th c. German woodcuts) and the life of Hercules (c. 1634). During recent restoration work similar sgraffito decoration from the 16th and early 17th c. depicting Biblical scenes was discovered on the side facing the square (including Joseph's Flight from the Wife of Putiphar).

Schwarzenberg Palace and Museum of Military History
(Schwarzenberg-palota, Vojenské muzeum)

The Schwarzenberg Palace, which now houses the Swiss Embassy as well as the Museum of Military History, ranks with the Castle and the Archbishop's Palace as a dominant feature in the panorama of the Hradčany. With its richly decorated gables, Lombardy-style protruding cornices, sgraffito ornamentation in imitation of faceted masonry (added in 1567 using North Italian and in particular Venetian graphic designs, reconstructed in the 19th c. and then again in the middle of the 20th c. and representations of Classical divinities and allegorical figures, it is a very characteristic example of the Renaissance architecture of northern Europe.

Opening times
May–Oct., Tues.–Sun.
9 a.m.–3.30 p.m.

The palace, on the south side of the square, was created in 1800–10 by the reconstruction in Empire style for Archbishop Salm – hence the initial S over the doorway – of two Renaissance mansions; the architect was F. Paviček.

The Museum is in the right-hand part of the building (1545–63, by Agostino Galli). It illustrates the development of the art of war down to 1918. Tempera frescoes (c. 1580) by an unknown artist adorn the ceiling in the main hall on the second floor. They depict personifications and allegorical figures from the famous poems of Homer (The Judgment of Paris, Helen's Abduction, Scenes from the Trojan Wars and the Flight of Aeneas from the burning Troy). The Phaeton Saga may be seen in the adjacent room. Two other rooms contain pictures of Chronos and Persephone as well as Jupiter amd Juno. The courtyard contains cannon and naval guns of the 16th to 20th c. as well as examples of prehistoric weapons and the great variety of armaments used by European armies over the centuries. There are also extensive displays of

Schwarzenberg Palace

army uniforms of all ranks and many different countries, a collection of medals, flags and banners, maps and plans of important battles.

Jewish Town Hall

See Josefov

Josefov (Joseph's Town, Jews' Town)   D4

Location
Jáchymova,
Staré Město, Praha 1

Metro
Staroměstská

Buses
133, 144, 156, 197

Tram
17

In the northern part of the Old Town is the old Jewish quarter which came to be known as Joseph's Town, after Emperor Joseph II. It is considered to be the oldest and most important Jewish community in the entire western world. A number of buildings in this quarter, including the Town Hall, the Synagogues and the cemetery, are open to the public as part of the State Jewish Museum.

History
Legend has it that a group of refugees appeared to Princess Libuče in a dream and promised to bring many blessings upon her land. Mindful of her mother's dream, Nezamyl accepted a group of refugees from Lithuania and allowed them to settle on the left bank of the Vltava. When this place below Bořivoj was no longer large enough to contain them all, the prince allocated the Jews a new dwelling place on the right bank of the river from which the Jewish Quarter later grew.
The first written mention of Prague, *c.* 965 by the Jewish merchant Ibrahim Ibn Jakub refers to the fact that Jews were prob-

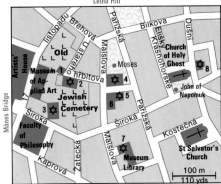

Josefov/
Joseph's Town
(Jews' Town)

1 Ceremonial House (Obřadní síň)
2 Klaus Synagogue (Klausova synagóga)
3 Pinkas Synagogue (Pinkasova synagóga)
4 Old-New Synagogue (Staronová synagóga)
5 High Synagogue (Vysoká synagóga)
6 Jewish Town Hall (Židovská radnice)
7 Maisl Synagogue (Maislová synagóga)
8 Spanish Synagogue (Španělská synagóga)

Old Town Square

ably settled in Prague, in the Lesser Quarter and below the Vyšehrad, before the 10th c. They came mainly as merchants because it was forbidden for them to be farmers or craftsmen. From the end of the 12th c. they lived in a separate quarter enclosed by a wall with seven gates, in the vicinity of what is now the Old-New Synagogue. This came about following the decision of the Third Lateran Council (1179) that the Jews should be separated from the dwellings of the Christians by a fence, a wall or a moat. This ghetto was considerably extended in the 17th c. In the reigns of the Habsburg rulers Maximilian and Rudolf II there were at times more than 7000 Jews crowded into an area of increasingly cramped and sunless lanes.

Rudolf II's Minister of Finance, Mordechaj Markus Maisl, had the streets of the ghetto paved, built the Jewish Town Hall and Maisl Synagogue and established the Old Jewish Cemetery which now attracts so many visitors.

In 1512 the first Hebrew printing-press in Central Europe was set up in this quarter. Schools, synagogues and public baths, an infirmary, a poor-house and a burial fraternity were established. A notable inhabitant of the ghetto (1573 onwards) was Rabbi Löw, a theologian and student of the Cabbala who features in many legends and is said to have created a golem (an artificial human being). The Jewish quarter, however, was frequently ravaged by fires (1378, 1754) and by pogroms. In 1389, for instance, more than 4000 Jews were killed in the ghetto and plundering and violently enforced eviction devastated the Jewish Quarter several times through the centuries.

After the Empress Maria Theresa (1743–80) decreed in December 1744 that all Jews be expelled from Bohemia, the last inhabitants of the ghetto lost their homes in the spring of 1745. Just three years later, the effect on the economy and pressure from foreign powers brought about a dispensation and the return of the Jews. Emperor Joseph II (1780–90) haad the walls of the ghetto pulled down and made the old Jews' Town the fifth ward of the town of Prague. Thereafter it became known as Joseph's Town (Josefov). It was not until after the Revolution of 1848 that Jews received civil rights. The freedom of residence granted to them led to considerable changes in the population structure of Joseph's Town; whilst the more wealthy were moving to other parts of the town, the poorer sector, both Christian and Jewish, remained

behind in the old houses and one-family dwellings with no water-supply or drainage.

The slum clearance which began in 1893 left only the buildings of historical interest still standing, and these now form the State Jewish Museum.

Until 1939 many of Prague's Jews were German-speaking, and writers such as Franz Kafka, Franz Werfel, Egon Erwin Kisch and Max Brod made a major contribution to German literature in the years between the two world wars. The Nazi occupation of 1939–45, however, dealt a mortal blow to the city's lively Jewish community. During the Second World War 90 per cent of the Jewish population of Bohemia and Moravia were killed: their names are inscribed on the "Memorial of the 77,297" set up in the Pinkas Synagogue to commemorate the victims of the Nazi terror. The memorial, recently damaged by flood water will, after reconstruction and restoration, replace the corresponding wall-mounted panels inside the Pinkas Synagogue which also bear the names of the victims.

**State Jewish Museum (Státní Židovské muzeum)

The State Jewish Museum (under State control since 1950) is not confined to material from the old Jews' Town of Prague. While seeking to exterminate the Jews the Nazis set out to develop the existing Jewish Museum, then very small, into an "Exotic Museum of an extinct race", and during the period of Nazi occupation the collection grew to a total of almost 200,000 items, with synagogues in Bohemia and Moravia and elsewhere in Europe making compulsory contributions to this unique documentation of Jewish life and faith.

Opening times
Apr.–Oct., Sun.–Fri.
9 a.m.–5 p.m.;
Nov.–Mar., Sun.–Fri.
9 a.m.–4.30 p.m.

Conducted tours
On request

Jewish Town Hall (Židovská radnice)

The Jewish Town Hall is now the headquarters of the Jewish Community of Prague and of the Council of Jewish Communities in Czechoslovakia.

The Town Hall, presented to the Jewish community by Mordechaj Markus Maisl, Burgomaster of the Jews' Town in the time of Rudolf II, was built in 1586 by Pankraz Roder in Renaissance style, and in 1765 was remodelled in Baroque style by Josef Schlesinger. The extension at the south end was added in the first decade of the 20th c. On the north gable, under the wooden tower, is a clock with Hebrew figures; the hands go anti-clockwise, since Hebrew is read from right to left.

Maislova 18

High Synagogue (Vysoká synagóga)

The High Synagogue, originally belonging to the Jewish Town Hall, is used by the State Jewish Museum for special exhibitions of ritual Jewish textiles, modern manuscripts and silver utensils. The fundamental part of the textile collection was previously housed in the Spanish Synagogue but a new, representative exhibition, designed by Zdeněk Rossmann was set up in 1981/82. The items on display are presented in chronological order dating from the end of the 16th c. to the 1930s. Bohemian embroideries and ritual textiles predominate but in addition to these, there are also fabrics imported into Bohemia from Central Europe and the Near East. The first room contains items with a domestic background (Sabbath and Passover covers, bridal caps and wedding

Cervená 4

◄ *Old-New Synagogue and Jewish Town Hall*

goblets) whilst the second has a collection of Torah wrappers with examples dating from *c.* 1570. This wrapper was prepared in the house on the birth of a son and consecrated when the child made his first visit to the synagogue. There he was swathed in the Pentateuch, or Torah, and was only allowed to pass out of it at his coming-of-age ceremony at 13. Four pieces of linen stitched together form the Torah which is decorated with inscriptions and, ornaments symbolising good wishes for the boy's future. The most recent exhibits (19th c.) are the cushions and covers for the lectern in the Torah shrine. The robe of Schelomo Molcho (*c.* 1500–30) from Portugal is of special interest. The third room is devoted to screens, draperies and Torah mantles which mainly originated in Prague. Among the most notable pieces are the appliquéd Torah screens decorated in Renaissance style which include Mordachaj Maisl's screen (1593) made for the synagogue founded by him. Splendid embroideries from the 17th/18th c. are identified by their Baroque embellishment (including the screen from Brunn, 1697; Torah mantle from the Klaus Synagogue, 1696; drapery from the Pinkas Synagogue, 1786). National synagogue textiles from Bohemian and Moravian Jewish communities may be seen in Room IV (including some from Brünn, 1837). The High Synagogue was built by Pankraz Roder in 1568 as a square hall in Renaissance style. In the 19th c. it was separated from the Town Hall and given its own entrance from the street and staircase. The main room on the first floor of the Town Hall – hence "High Synagogue" – was enlarged in the 17th c. and remodelled in neo-Renaissance style in the 19th. With its beautiful stellar vaulting this room – a characteristic example of Jewish religious architecture – is in sharp contrast to the plain exterior of the synagogue.

Old-New Synagogue (Staronová synagóga)

Facing the High Synagogue is the Old-New Synagogue (closed from 4 p.m. on Fridays). It was originally known as the "New" or "Great" Synagogue but received its present name after another synagogue was built in the Jewish Town in the 16th c. It is the only synagogue of its period in Europe which is still in use for worship.

The oldest part is the Early Gothic south hall, originally the main hall of the synagogue, to which a two-aisled hall in Cistercian Gothic style was added in the 13th c. The five-ribbed vaulting is unique in Bohemian architecture. The brick gable of the synagogue dates from the 15th c., the original timber gable having been destroyed in the great fire of 1338. The women's galleries were completed in the 17th and 18th c. (the main hall being reserved for men). The large flag was presented by the Emperor Ferdinand III in 1648 in recognition of the Jewish community's contribution to the fight against the Swedes during the Thirty Years' War. This red flag with a star and cap has served ever since as the official banner of the Jews of Prague. It is even represented on the capital's historic coat-of-arms being the third flag from the right next to the Bohemian lions. This makes Prague the only city in the world to carry the Jewish colours in its coat-of-arms. In a Torah shrine on the east side of the hall is a parchment scroll of the Pentateuch, the five Books of Moses. In the centre of the hall stands the pulpit, set apart by a 15th c. screen. The Hebrew inscriptions on the walls refer to a renovation of the synagogue in 1618. Further reconstruction took place in 1883 and 1966. Legend has it that the remains of the Golem are to be found in the loft of the Old-New Synagogue. The famous Prague legend was created

Old-New Synagogue: Torah shrine

Klaus Synagogue

in 1580 A.D., 5340 according to Jewish chronology, by Rabbi Löw in order to protect the Jews of Prague from persecution. The history of the Golem is recorded in literary works not only by Czech but also by German writers such as Gustav Meyrink (1868–1932) in his novel "The Golem" and the reporter Egon Erwin Kisch (1885–1948) "On the tracks of the Golem" in his work "Powder Tower".
In the gardens adjoining the synagogue is a statue of Moses by František Bílek (1872–1941).

Klaus Synagogue (Klausová synagóga)

From the Old-New Synagogue the street called U starého hřbitova (By the Old Cemetery) leads to the Klaus Synagogue.
The Klaus Synagogue, a Baroque building erected in 1694 and remodelled externally in 1884, houses an exhibition of Hebrew manuscripts and old prints. Rabbi Löw ben Bezalel, one of the leading Jewish philosophers of the 17th c., taught in this synagogue.

Ceremonial Hall

To the right of the entrance to the Old Jewish Cemetery is the neo-Romanesque Ceremonial Hall with a small tower from 1906. The hall is used for special exhibitions but on permanent display here are children's drawings, school books, diaries, letters and poems from the Theresienstadt (Terezín) concentration camp.

Old Jewish Cemetery (Starý źidovsky hřbitov)

The Old Jewish Cemetery ranks with the Old-New Synagogue as one of the most important features of the old Jews' Town – and, as some would have it, one of the "ten most interesting sights in the world". It was established in the first half of the 15th c. and remained in use until 1787. Under its elder trees are no fewer than 20,000 gravestones. The restricted area of the cemetery was inadequate for the large numbers of burials, and additional earth had to be brought in to accommodate more graves. The result is that in places there are anything up to nine superimposed layers of burials. Hence the extraordinary accumulation of gravestones, huddled together in picturesque confusion.
The Hebrew inscriptions on the gravestones give the name of the dead man and of his father (in the case of women the husband's name as well), together with the dates of death and burial and a recital (sometimes in verse) of the dead person's good works. The reliefs on the gravestones frequently symbolise his name (e.g. a stag if his name is Hirsch, a bear if his name is Bär) or occupation (a doctor's instruments, a tailor's scissors), or may depict other symbols such as hands in the attitude of benediction, sacred vessels (for members of priestly families), grapes (for members of the tribe of Israel), crowns or pine-cones.
The oldest gravestone is that of the scholar and poet Avigdor Karo (d. 1439), who lived through the pogrom of 1389 and wrote an elegy on it. The most recent grave is that of Moses Beck (d. 1787) but no further burials have taken place here since that date. A Late-Renaissance sarcophagus with carved lions and writing tablets framed by arches marks the tomb of the learned Rabbi Jehuda Löw ben Bezalel (1609).

Tomb of Rabbi Löw

Bassevi gravestone

Other notable graves are those of Mordechaj Markus Maisl, Burgomaster of the Jews' Town (d. 1601), the historian and astronomer David Gans (d. 1613), the learned Joseph Schelomo Delmedigo (d. 1655) and the scholar and bibliophile David Oppenheim (d. 1736). One of the most beautiful and most richly decorated is that of Heudele Bassevi (d. 1628), wife of Jakob Bassevi, Treasurer of the Waldstein family and the first Jew from Prague to be raised to the peerage.

The pebbles accumulated on many of the graves are deposited by friends and relations of the dead person in token of respect and esteem. Pious visitors throw notes into the tomb of Rabbi Löw expressing wishes which they ask the wonder-working Rabbi to fulfil.

Pinkas Synagogue (Pinkasova synagóga)

The Pinkas Synagogue (at present in course of reconstruction: not open to the public until the 90s), on the south side of the Old Jewish Cemetery, was originally established in a house which the Horowitz family, the leading family of the Jewish community, had bought from Rabbi Pinkas in the 14th c. In 1535 Salman Munka Horowitz built a synagogue in Late Gothic style, which Juda Goldschmied de Herz remodelled in the style of the Late Renaissance in 1625, adding a women's gallery, a vestibule and a meeting-room.

Closed at the present time

Architecturally the Pinkas Synagogue is the finest of the Prague synagogues. Archaeological excavation has confirmed that the building dates from the 11th or 12th c. and that there was a ritual bath here.

After the Nazi persecutions of 1939–45, which practically wiped out the Jewish population, the "Memorial of the 77,297" was erected (1950–58), listing the names of all those who were killed. Because of damage caused to the memorial by an underground spring in the old ritual bath, the names of the victims are now recorded on several wall-mounted tablets in the synagogue.

Maisl Synagogue (Maislova synagóga)

This synagogue was founded by Mordechaj Markus Maisl, Burgomaster of the Jews' Town in the time of Rudolf II, as a family place of prayer; the architects were Joseph Wahl and Juda Goldschmied. It is a three-aisled Renaissance building (1590–92) with 20 columns supporting the roof. The building was restored in Baroque style following a fire in 1689 and then, between 1893 and 1905, it was remodelled in neo-Gothic style by Alfred Grotte. The Maisl Synagogue, whose art treasures will not be on view to the public until possibly 1991 because of restoration work which is in progress, houses an exhibition of silver from 153 Bohemian synagogues and private houses. The collection includes ornaments for Torah scrolls, breastplates, crowns, Torah pointers, cult objects, including spice-boxes and etrog boxes, cups, goblets and candlesticks. This is a magnificent display of Bohemian silversmiths' work of the 17th to 19th c. and work by superb Nuremberg, Augsburg and Viennese craftsmen of the Baroque and Rococo periods.

Spanish Synagogue (Španělská synagóga)

Dušni 12

The building occupies the site of Prague's first synagogue (12th c.), known as the Old School which was destroyed.

The name of the synagogue comes from a group of Jews who fled from the Inquisition in Spain and settled in Prague. Through the centuries the synagogue was burnt down several times but was always rebuilt. It was given its present Moorish-style form, with an imposing dome and fine Roman arches between 1882 and 1893 by V. I. Ullmann. Inside, the stucco decoration imitated from the Alhambra of Granada was added between 1882 and 1893. Restoration work being carried out on the Spanish Synagogue will also last into the 1990s.

Kampa Island　　　　　　　　　　　　　　　　　　　　E3

Location
Malá Strana, Praha 1

Metro
Malostranská

Trams
9, 12, 18, 22

Kampa Island is the swathe of green which extends along the left bank of the Vltava from 1st May Bridge to the Mánes Bridge, separated from the Lesser Quarter by the idyllic (but in the past also dangerous) arm of the river known as the Čertovka or Devil's Brook. The western end of the Charles Bridge crosses the island, and some of its houses have foundations built against the piers of the original Judith Bridge. To the north of the Charles Bridge the Čertovka flows between two lines of houses which are often referred to as the "Venice of Prague" (Pražské Benátky).

In earlier times the island was marshy, with some land laid out as gardens, and the first houses were not built until the 15th c. The Čertovka then served to drive mill-wheels – still to be seen at the Charles Bridge (see entry) and the bridge leading to Grand Prior's Square (Velkopřevorské náměstí). The old Sova mill is to be converted into a luxury hotel.

The attractions of Kampa Island are its pottery markets, the pleasant walks it affords and the fine views of the Vltava with the Střelecký ostrov (Marksman's Island), the Charles Bridge, the Old

Kampa Island

Town and the garden fronts of some of the palaces in the Lesser Quarter. The large park in this area was created by amalgamating the old palace gardens.

Just by the Charles Bridge is a Late Gothic figure of Roland, reconstructed by L. Šimek in 1884, which once marked the boundary between the Lesser Quarter, governed by the Magdeburg legal code, and the Old Town, governed by the Nuremberg code.

Karlovo náměstí

See Charles Square

Karlovo ulice

See Charles Street

Karlův most

See Charles Bridge

Karlštejn Castle (Hrad Karlštejn)

Above the little wine-producing town of Karlštejn (pop. 1200) rears Karlštejn Castle (Hrad Karlštejn, formerly Karlův Týn), the

Distance
28 km (17 miles) SW

Karlštejn Castle

Karlštejn Castle

Opening times
Open-air performances:
May–Aug./Sept., Sat. and Sun.
7–11 p.m.

most celebrated of the Bohemian castles. It lies to the north at the top of a limestone cliff (319 m/1047 ft above sea-level) in the lateral valley of the Beraun (Berounka) and may be reached on foot (2 km) from the car-park just outside the village – in summer the buses operate a shuttle-service. The castle, a national monument, was probably designed by the French architect Matthias of Arras. It was built by the Emperor Charles IV between 1348 and 1357 as a place of safety in which the Crown Jewels of the Holy Roman Empire, the royal insignia of Bohemia and numerous relics could be kept. It was partially rebuilt in the 15th and 16th c. and was restored, with numerous alterations, between 1888 and 1904 (architects Friedrich Schmidt and Josef Mocker).

Castellan's Courtyard
The Castellan's Courtyard (Purkrabský dvůr) is entered from the north through two gate-buildings which lie roughly 100 m (328 ft) apart. The conducted tour starts from the courtyard which has been fitted out for use as an open-air theatre.

Castellan's Lodging
The Castellan's Lodging (Purkrabství) is on the south side of the courtyard. The lower parts of the 4-storeyed building date from the 15th c.

Well Tower (Studniční věž)
At the westernmost tip of the castle are the old domestic offices and the large Well-Tower, with a 90 m (295 ft) deep well and large well-wheel.

Imperial Palace (Císařský palác)
From the Castellan's Courtyard a large gateway leads into the narrow Inner Ward (Hradní nádvoří), on the right of which is the

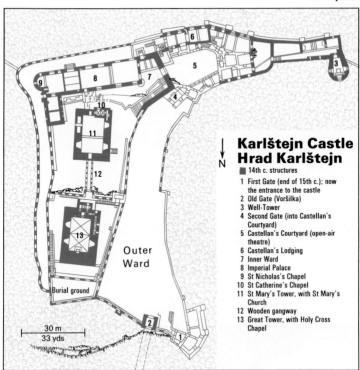

Karlštejn Castle
Hrad Karlštejn

■ 14th c. structures

1 First Gate (end of 15th c.); now the entrance to the castle
2 Old Gate (Voršilka)
3 Well-Tower
4 Second Gate (into Castellan's Courtyard)
5 Castellan's Courtyard (open-air theatre)
6 Castellan's Lodging
7 Inner Ward
8 Imperial Palace
9 St Nicholas's Chapel
10 St Catherine's Chapel
11 St Mary's Tower, with St Mary's Church
12 Wooden gangway
13 Great Tower, with Holy Cross Chapel

Imperial Palace (Císařský palác). A staircase, far right, leads up to the first floor. In the first room are casts of the fine busts from the triforium of St Vitus's Cathedral (see Hradčany) and illustrative material (mostly photographs) on Charles IV and Prague as it was in his reign. In the second room are remains of stained-glass windows from the Holy Cross Chapel, pictures (including a diptych from the Palmatius Church in Budňahy; 15th c.), sculpture and architectural models illustrating the history of the castle.

Of the former Imperial apartments on the second floor only Charles IV's study, his bedroom with an altar-piece by Tommaso da Modena and the so-called Imperial Room with fine panelling, have been preserved. The half-timbered top storey with the women's quarters was replaced during the restoration work by a wooden wall-walk.

St Nicholas's Chapel
This chapel at the east end of the Imperial Palace is not open to the public.

St Mary's Tower
This tower (Mariánská věž) stands immediately north of the Imperial Palace but is at present closed to the public owing to restoration work. On the second floor (reached by a staircase in the thickness of the wall) is the Capitular Church of St Mary,

101

Karlštejn Castle

Opening times
Open-air performances:
May–Aug./Sept., Sat. and Sun.
7–11 p.m.

which has a painted timber ceiling and wall-paintings (some dating from the 14th c.) of scenes from the Apocalypse and portraits of Charles IV.

In the south-west corner of the tower is the vaulted Chapel of St Catherine. This originally had wall-paintings, which Charles IV caused to be replaced by a facing of rare stones. Above the entrance there is a painting of the Emperor, Charles IV and his wife Anna, and an original image of the Virgin Mary may be seen in the altar niche.

Great Tower or Keep (Velká věž)
On the highest point of the castle rock is the massive Great Tower, 37 m (121 ft) high, which is connected with St Mary's Tower by a wooden gangway (originally there was a drawbridge).

Holy Cross Chapel (Kaple svatého Kříže)
Closed to the public at the present time, this chapel, on the second floor, was consecrated about 1360. A gilded iron screen divides it into two parts. Its low vaulted ceiling is entirely covered with gilding and set with glass stars giving the illusion of a firmament. On the walls, above the racks of candle-holders (for 1330 candles), are more than 2200 semi-precious stones set in gilded plaster and 128 painted wooden panels symbolising heavenly armies (1348–67) by Master Theoderich (behind which relics were originally preserved). Behind the altar is a niche in which the German Imperial Crown Jewels (now in the Treasury of the Hofburg in Vienna) and later also the Bohemian royal insignia (now in the Crown Chamber in St Vitus's Cathedral: see Hradčany) were kept. A large memorial mass is held here each year in the early evening of 29 November to celebrate the anniversary of the death of Charles IV.

Chapel of St Catherine

Holy Cross Chapel

Kinsky Palace (Palác Kinských), E4
Graphic Collection of National Gallery (Grafická sbírka Národní galerie)

The Kinsky Palace on the east side of the Old Town Square, now houses the National Gallery's Collection of Graphic Art, with a fine collection on permanent display and periodic special exhibitions.

The palace is a Late Baroque building with fine Rococo features, erected on the foundations of earlier Romanesque buildings. Commissioned by J. E. Goltz in 1755, and originally designed by Kilian Ignaz Dientzenhofer, it was completed by Anselmo Lurago (1755–65). Three years later, it was bought by Prince Rudolf Kinsky. The main front is almost neo-Classical in style, with its rich stucco decoration by C. G. Bossis, projecting bays framed in pilasters and two pediments. Across the front, borne on the columns flanking the two entrances, is a balcony (from which speeches were made on special occasions such as Clement Gottwald's historic speech to the Czechoslovak people on 21 February 1948). Fronting the attic storey are eight mythological figures (four erect, four reclining) by Ignaz Platzer the Elder. Rococo influence is shown by the delicately articulated window-framings with the characteristic rocaille motif. The three side wings in Empire style are later additions.

Immediately next to the Kinsky Palace is "Stone Bell House" (At the sign of the Stone Bell). The front with its soaring Gothic windows was covered by a neo-Baroque façade (1899) until repair work was carried out in the 1960s. The history of the construction of the house goes back to the second half of the 13th c. and records show that it was named "At the sign of the Stone

Location
Staroměstské náměstí (Old Town Square) 12, Staré Město, Praha 1 (Pedestrian zone)

Metro
Staroměstská
Můstek

Stone Bell House (U kamenného zvonu)

Kinsky Palace and "stone bell" house

Bell'' for the first time in 1747. It was probably built by Queen Elisabeth, the wife of John of Luxembourg. Structural changes were made at the end of the 15th c. and a conversion, after 1685, ensured that scarcely anything remained which bore any resemblance to the palace that had once been the home of kings. The building recovered its Gothic form in 1987 when extensive reconstruction work was carried out. Today the Prague Gallery holds exhibitions here of retrospective and modern art as well as concerts and recitals.

Klaus Synagogue

See Josefov

*Knights of the Cross Square (Křížovnické náměstí) E4

Location
Křížovnické náměstí
Staré Město, Praha 1

Metro
Staroměstská

Trams
17, 18

St Salvator's Church
(Kostel svatého Salvátora)

Knights of the Cross Square, one of Prague's most picturesque squares, came into being in the 16th c. at the end of the Charles Bridge (see entry). The traditional coronation procession of the Bohemian kings passed through the square. On its east side is St Salvator's Church, on its north side the Knights' Church of St Francis.
The Military Order of the Knights of the Cross with the Red Star developed out of a brotherhood of hospitallers in the period of the Crusades; it was mainly active in Silesia, Bohemia and Moravia.

St Salvator's Church, on the east side of Knights of the Cross Square, was originally a Jesuit church within the Clementinum

Holy Cross Church of St Francis

St Salvator's Church

(see entry) complex. It was built in Renaissance style between 1578 and 1601. The Baroque porch was added in 1638–59 by Carlo Lurago and Francesco Caratti; the vases and statues of saints were the work of Johann Georg Bendl (1659). The towers were built in 1714 (architect F. M. Kaňka). The ceiling-painting of the four quarters of the world is by K. Kovár (1748).

This Baroque church, which belonged to the Order of the Knights of the Cross, was built by Jean-Baptiste Mathey in 1679–89 on the foundations of an Early Gothic church of which some remains survive underground. It has a splendid dome. The façade, in the manner of the French pre-Classical school, is decorated with statues of angels and the patron saints of Bohemia. The figures of the Virgin and St John of Nepomuk in front of the entrance are from the workshop of M. W. Jäckel (1722).

Church of St Francis Seraphicus (Kostel svatého Františka Serafinského)

Notable features of the richly decorated and furnished interior are the large fresco of the Last Judgment (by V. V. Reiner, 1722) in the dome, the altar-piece by J. K. Liška and M. L. Willmann and a 15th c. Gothic Madonna beside the side altar.
Outside the church is the Vintners' Column, by Johann Georg Bendl, with a statue of St Wenceslas (1676).

Between St Francis's Church and the Old Town Bridge Tower (see Charles Bridge) is a cast-iron statue of Charles IV, erected in 1848 on the 500th anniversary of the establishment of Prague University.

Monument to Charles IV

The Koněprusy Stalactite Caves

The Koněprusy Caves (Koněpřuské jeskyně) were not discovered until 1950. The caves with bizarre shapes formed by the growth of stalactites and stalagmites are connected with the limestone formations of the Bohemian karst (Český kras) and are the largest in the country. Human and animal remains were uncovered when the site was being excavated. The work-shops of 15th c. forgers have been rebuilt and a visit would be worthwhile.

Location
45 km (29 miles)
southwest of Prague

Opening times
April, Sept., Oct. daily
8 a.m.–3 p.m. (Sept. Sat. &
Sun. to 4 p.m.); May–Aug.
Mon.–Fri. 8 a.m.–4 p.m., Sat.
& Sun. to 5 p.m.

*Konopiště Castle (Zámek Konopiště; field of hemp)

This castle near Benešov u Prahy (373 m/1224 ft above sea-level) is a favourite place for the people of Prague to visit. There is a car-park beneath the castle some 2 km from the town. The castle itself stands in an elevated position and a pleasant 10-minute walk takes the visitor to the entrance at the east tower. The courtyard is entered through a Baroque gate (F. M. Kaňka) with sculptures dating back to 1725 by M. B. Braun. The original Gothic fortification (13th/14th c.), based on a French design, was reconstructed at the beginning of the 16th c. in Late Gothic style and extended in the early 17th c. by the addition of a Renaissance palace. Baroque elements were introduced in the 18th c. Konopiště was bought in 1887 by the Austrian Archduke and heir to the throne, Franz Ferdinand d'Este who was assassinated in Sarajevo in 1914. He turned the castle into a sumptuous palace, the design of which was drawn up by Josef Mocker. The priceless interior decoration including the objets d'art contained in the rather unusual St George's Museum, which has numerous paintings, sculptures and similar exhibits all pertaining to St George, stem

Location
44 km (28 miles)
southwest of Prague

Opening times
Tues.–Sun. May–Aug.
9 a.m.–6 p.m.;
April, Sept., Oct.
9 a.m.–4 p.m.

from the same period. The vast collection of weapons (mainly from the estate of the d'Este family of Modena) with almost 5000 pieces (including armour from the 15th and 16th c., precious swords and muskets) is the most impresssive in Europe. The collection of hunting trophies and other memorabilia (paintings, porcelain, Gobelins, etc.) from the Archduke's travels around the world is extraordinarily large.

Křivoklát Castle (Pürglitz Castle)

Location
40 km (25 miles)
west of Prague

Opening times
Tues.–Sun. May–Aug.
9 a.m.–12 noon, 1 p.m.–6
p.m.; April, Sept, Oct. 9
a.m.–4 p.m.

Rising amid dense woodland on a rocky ledge above the Rakonitz (Rakovnický potok), a tributary of the Berounka, is the imposing Křivoklát Castle, one of the oldest preserved fortifications in Bohemia. The royal castle was first mentioned in records at the beginning of the 12th c., extended and fortified at the end of the 14th c. and one hundred years' later re-built in Gothic style. The building was extensively restored in c. 1920.

Křivoklát became the royal residence of the Přemysliden dynasty during the reign of Ottokar II (1252–78). Charles IV and Wenceslas IV enjoyed hunting here but during the reign of the Habsburgs the castle lost its significance and later changed hands on more than one occasion.

Remains of the Late Romanesque Palatinate and the enclosed palace from the time of Ottokar II are preserved. The sumptuous winged altar with carved figures of the apostles and saints is a notable feature of the High-Gothic castle chapel.

Classical concerts and theatre performances take place in the castle grounds during the summer months.

The vast woods of Křivoklát (63,000 ha), between Beroun and Rakovník (the traditional centre of hop-culture) make up one of the three regions of Czechoslovakia which were included in UNESCO's international MAB Programme (Man and the Biosphere). This involves the promotion of agricultural management practices aimed at protecting the environment in addition to research projects which comprise measures to protect the environment generally.

**Kutná Hora (Kuttenberg)

Situation
68 km (41 miles)
east of Prague

This central Bohemian town (273 m; 903 ft above sea-level) rose to fame and wealth on account of the abundant deposits of silver discovered in the area. Overlooking the valley of the river Vrchlice, the picturesque mining town which is scheduled as an ancient monument, was the largest town in Bohemia during the Middle Ages. From time to time it was also a royal residence. The town gained its importance from the silver mines worked here between the 13th and 18th c. which provided the base for the Prague silver Groschen produced in the Kuttenberg Mint from 1300. This was the most stable and best known coin in Bohemia during the Middle Ages. One of the most significant dates in the town's history is 1409, when the governing body of Prague's Charles University issued the amended Kuttenberg Decree granting favour to the Czechs and giving rise to the emigration to Leipzig of German professors and students.

The famous Baroque artist, Johann Peter Brandl died in Kuttenberg in 1735 (born in Prague in 1668) after spending the last years of his life in the town. In addition, the founder of the modern Czech theatre, Josef Kajetán Tyl (1808–56), who also wrote the Czech national anthem, was born here.

Sights

The unparalleled masterpieces of Gothic architecture reflect the town's period of prosperity. In the historic town-centre, focused on Palackého náměstí (Palacký Square), which is fringed by Renaissance houses, the visitor should look for the so-called "Stone House" (Kamenný dům) in Náměstí Privního máje (May 1st Square). The 15th c. building with bay-windows and sculptural decoration today houses the municipal government and Museum.

Stone House

The Baroque Ursuline Convent (1733–43), designed by Kilian Ignaz Dientzenhofer, may be seen to the north-east on the Třida Jiřiho z Poděbrady (George of Podiebrad Street).

The Baroque Church of St John of Nepomuk (Kostel svatého Jana Nepomuckého), built by F. M. Kaňka, (1734–54) is in the Husova třida (Hus Street) which begins on the western side of the Šultysovo náměstí (Sultys Square). The ceiling-paintings in the interior of the church were completed in 1752.

Church of St John of Nepomuk

To the west, a little way up the hill, Hus Street opens out on to the Rejskovo náměstí (Rejsek Square) in the centre of which stands a Late Gothic dodecagonal stone fountain (Kamenná kašna; 1493–97. The marble house (U Marmorů) on the south-west corner of the square has a fine Renaissance portal.

The downhill walk from Komenského náměstí (Comenius Square) to the U Vlašského dvora Square (Court of the Italians) where the Gothic Church of St John (Chrám svatého Jakuba) stands, is a little more easy-going. The church was built between 1330 and 1420 and has a bell-tower which is 82 m (238 ft) high. Its rich interior decoration includes a splendid high altar (1678) and notable paintings, among them works by J. P. Brandl. Facing the church, to the south, is the Archdeaconry which was once the home of the Master of the Mint.

Church of St John

Adjoining the square on the east is the Court of the Italians (Vlašský dvůr) (c. 1300), the old mint, which was named after the first coin-makers who came from Florence and which later served for a long time as a royal palace. The oriel of the Gothic Latin Chapel may be seen in the attractive courtyard approached from the east. Inside the chapel, there is a colourful carved altar commemorating the Death of Our Lady. The museum in the building has a display of old workshops used by the coin-makers as well as pictures by the Kuttenberg artist, Felix Jenewein (1857–1905). There is a good view from the terraces on the south side of the Court of the Italians, particularly of St Barbara's Church.

Court of the Italians

To the south-west of St John's Church, on the Barborská ulice (Barbara Street) stands the castle (Hrádek) built in the 15th c. on the site of a 14th c. wooden fortification. Shortly after completion it became the second Mint and now houses the Miners' Museum. There are some elaborate wall-paintings in the Gothic Hall of Knights. The two oriels in the chapel are also worth seeing.

Castle

A little further along Barbara Street is the massive Jesuit College (Jesuitska kolej) which for a long time has served as a barracks.

Jesuit College

At the southern end of Barbara Street is the Gothich Church of St Barbara (Chram svaté Barbory). It was begun in 1388 by Peter Parler, continued later by Matthias Rejsek and Benedikt Reid and finally completed in 1585. Restoration and completion of the façade followed between 1884 and 1905. The church building with five aisles, choir gallery, chapel cornice and numerous buttresses has a roof formed from three canopy-like pyramids which

**Church of St Barbara

tower above the main aisle. The many coats-of-arms on the reticulated vaulting in the nave, which is both impressive and spacious, reflect a Renaissance influence. There are beautiful Gothic choir stalls in the north aisle and exposed frescoes in the choir chapel. In addition, there is a Renaissance pulpit dating from 1566 and a high altar carved in 1903 as well as pictures by K. Škreta and J. P. Brandl.

Church of Our Lady

A visit to the 15th c. Church of Our Lady (Matka Boží or Kostel Panny Mrie Na náměstí) on the eastern edge of the Old Town is worthwhile. It has splendid Gothic vaulting and a pulpit dating back to 1520.

Church of St Mary
Charnel House Chapel

In the part of the town known as Sedletz are the Gothic Baroque Church of St Mary (Chrám Panny Marie) and the Charnel House Chapel, so-named because it contains a vast collection of human bones as well as an altar, candlesticks, coats-of-arms, etc.

****Lesser Quarter** (Malá Strana)   D2/3 E1–3

Location
Malá Strana, Praha 1

Metro
Malostranská

Trams
12, 18, 22

The Lesser Quarter or Little Town is the name given to the district of Prague which was originally the "lesser town of Prague" founded in 1257, during the reign of Přemysl Ottokar II. In the reign of Charles IV the settlement grew; the churches of the Přemyslid period, some of them going back to the Romanesque period, were rebuilt, and the little town was enclosed within walls which extended as far as the Hunger Wall on Petřín Hill (see entry).

In the 15th and 16th c. the Lesser Quarter was devastated by three major fires, which led to further building and rebuilding. After the catastrophic fire of 1541 the town took on the aspect of a princely capital: wealthy nobles came here to live, and magnificent churches were built. After the Battle of the White Mountain in 1620, in which the forces of the Catholic League defeated the Protestants, many noble families from other parts of the Empire moved into Bohemia, and this gave a further impulse to the development of the Lesser Quarter. Whole streets and large areas of gardens were destroyed to make way for the splendid mansions of the Habsburg and Catholic nobility, such as the Nostitz Palace, the Waldstein Palace and the Buquoy Palace (see entries). Each noble family sought to establish a palace near the Hradčany (see entry) and if possible to surpass it in splendour.

St Nicholas's Church (see Lesser Quarter Square) with its dome became the dominant feature of the town, which otherwise spread out in horizontal lines. This was the decisive period in the building history of the Lesser Quarter, giving it its distinctly Baroque character.

The attraction of the quarter was enhanced by numbers of small squares including Waldstein Square, Grand Prior's Square (Velkopřevorské náměstí) and Maltese or Five Churches Square and by the gardens of monasteries and palaces, most of which are now open to the public.

From Bridge Lane (Mostecká) the coronation processions of the Bohemian kings crossed Lesser Quarter Square (see entry) and along what is now Neruda Street (Nerudova) to the Hradčany (see entry).

The Lesser Quarter ▶

Lesser Quarter Bridge Towers

See Charles Bridge

*Lesser Quarter Square (Malostranské náměstí)   E3

Location
Malá Strana, Praha 1

Metro
Malostranská

Lesser Quarter Square, the focal point of the Lesser Quarter since its earliest days, is divided into two smaller squares by the buildings around St Nicholas's Church.

In the lower square are such notable buildings as the Lesser Quarter Town Hall and the Late Baroque Kaiserstein Palace (U Petzoldu; c. 1700) on the east side and the Rococo house "At the Sign of the Stone Table" in the middle. Each square once had a fountain but the upper one was replaced in 1715 by a pest column with a statue of the Holy Trinity and the patron saints of Bohemia. The lower fountain was replaced in 1858 by a memorial (J. O. Mayer and F. Geiger) to the Austrian Marshall, J. W. Radetzky and this is now in the Lapidarium of the National Museum.

Kaiserstein Palace
(U Petzoldů)

Kaiserstein Palace (No. 23) which was reconstructed at the beginning of the 1980s and today belongs to the Chamber of Commerce occupies the site of two 15th c. Gothic houses. These were joined together in 1630 and converted in 1700 by Helfried Kaiserstein. The attics have allegorical paintings by O. Mosto of the four seasons and the family coat-of-arms is depicted above the centre window on the first floor. At the beginning of the 20th c., the celebrated prima donna, Ema Destinnová (Kittlová; 1878–1930), partner of the legendary E. Caruso, lived in the palace where there is a room especially dedicated to her.

Palace of Liechtenstein

The main feature of the upper square, known in the 17th c. as Latin Square and from 1874 also as St Stephen's Square, is the Palace of Liechtenstein. The building, now the Communist Party School, was erected in 1591 and restoration work was carried out right up until 1989. Its Classical façade was added in 1791.

Lesser Quarter Town Hall (Malostranská radnice)

The Town Hall of the Lesser Quarter has stood on this site since the late 15th c. The important negotiations on the Bohemian Confession were held here in 1575.

In its present form the Town Hall dates from the Late Renaissance (1617–22). The doorway with the town's coat-of-arms was added in 1660. Manuscripts dating from 1600, including the "Lesser Quarter Hymn Book" (1572) which is now preserved in the State Library, were discovered when building work was going on in the vaults. Nowadays, the Town Hall is used for cultural meetings.

**St Nicholas's Church (Chram svatého Mikuláše)

Formerly a Jesuit church, St Nicholas's Church occupies the site of an earlier Gothic church with the same dedication. Its construction was the work of three generations of the best Baroque architects of Prague. The mighty nave with its side chapels, galleries and vaulting was built by Christoph Dientzenhofer (1704–11), the choir with its dome by Kilian Ignaz Dientzenhofer (1737–52), and the tall tower which completed the church by Anselmo Lurago (1756). The masterpiece in Bohemian Baroque is the two-storeyed façade completed in 1710. The arms of the Count of Kolowrat above the main doorway as well as the statues of the early fathers of the church are from the workshops of Johann Friedrich Kohl.

St Nicholas's Church Chrám svatého Mikuláše

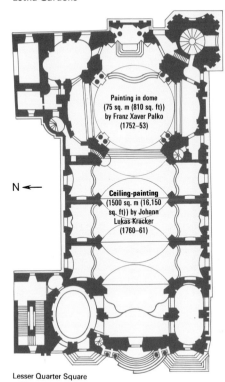

Painting in dome
(75 sq. m (810 sq. ft))
by Franz Xaver Palko
(1752–53)

Ceiling-painting
(1500 sq. m (16,150
sq. ft)) by Johann
Lukas Kracker
(1760–61)

N ←

Lesser Quarter Square

St Nicholas was a 4th-c. Bishop of Myra in Asia Minor who was regarded in medieval times as the patron saint of municipal administration and the guardian of justice.

The centre of the ceiling-paiting is dominated by the figure of the Saint, surrounded by angels, with his crosier in his left hand and his right hand raised in blessing. One scene shows a priest distributing phials of the wonder-working oil dispensed by the Saint. Another scene depicts Nicholas giving money to a poor man who is reduced to selling his daughter for gold.

A scene on the right-hand side refers to an occasion during the wars of the 14th c. when St Nicholas saved three Romans from execution by his intervention. On the left-hand side is a coastal landscape, alluding to the Saint's role as protector of seafarers and merchants. Coastal and harbour scenes were much favoured by the burghers of Prague during this period.

The sumptuously decorated interior, an example of High Baroque, achieves its awesome effect mainly through its superb frescoes. The ceiling-painting in the nave (by Johann Lukas Kracker, 1760–61) depicts scenes from the life of St Nicholas. The dome has representations of the Glorification of St Nicholas and the Last Judgment by Franz Xaver Palko (1752–53), who was also responsible, together with Joseph Hager, for the wall-paintings in the choir. The sculpture in the nave and choir and the figure of St Nicholas on the high altar are by Ignaz Platzer the Elder, the mighty organ was built by Thomas Schwarz in 1745.

The gilded pulpit by Richard and Peter Prachner (1765) is of artificial marble with Rococo decoration depicting allegorical paintings of Faith, Hope and Charity as well as the picture of "The Beheading of John the Baptist".

The side altar in the transept has paintings by Kracker of "The Visitation of the Virgin Mary" and the "Death of St Joseph" (1760). The other altar pictures and ceiling-paintings including those of Ignaz Raab, Francesco Solimena and Franz Xaver Balko are also worth seeing.

The cloisters of the old Minorite convent (13–14th c.) adjoin the church on the north side. Parts of them were remodelled in Baroque style in the 17th century.

St Nicholas's Church in the Lesser Quarter   *Painting in the dome by F. X. Balko*

*Letná Gardens (Letenské sady)   C/D3/5

Location
Na baště svatého Tomáše,
Hradčany, Praha 6
Konstelní,
Holešovice, Praha 7

Trams
1, 5, 8, 12, 17, 18, 20, 22, 23, 24

Hanau pavilion

North-east of the Hradčany (see entry), above the left bank of the Vltava, rises Letná Hill (Summer Hill) with its beautiful gardens. The coronation of Ottokar II took place here in 1261. In 1858 the area was taken over by the town and made into a park, designed by B. Wunscher and laid out by J. Bauerle. From Svatopluk Čech Bridge a stepped path (256 steps in all) leads up to the outlook platform on the massive base which once supported a 30 m (100 ft) high monument to Stalin (pulled down in 1962), and now affords panoramic views of Prague, Petřín Hill (see entry) and St Vitus's Cathedral (see Hradčany). Designed by Z. E. Fiala for the 1891 World Fair, the iron pavilion was cast in the foundries of Prince Hanau, from whom it took its name. In 1898 it came to Letná Hill where it was converted into an elegant restaurant, from which there are magnificent views.

Praha Expo 58

Near the east end of the gardens (formerly Belvedere Park) is the Praha Restaurant (F. Cubr, J. Hrubý and Z. Pokorný) which was originally part of the Czechoslovak section of the 1958 Brussels International Exhibition and was re-erected on its present site after the exhibition closed. The doorway is the work of J. Kodet, and the frescoes inside are by artists of the A. Fišárek School of the Academy of Graphic Arts. From here, too, there is a good view of Prague, although the Hradčany is obscured.

Stadion Sparta ČKD

The original football stadium was rebuilt and extended at the end of the 1960s. Additional pitches and a sports hall are contained within the grounds, and the stadium has accommodation for 40,000.

Lidice

Originally a small mining town, Lidice was razed to the ground by the Nazis on June 10th, 1942 in retaliation for the assassination of the Protector of the German Reich for Bohemia and Moravia, Reinhard Heydrich. All 173 of the men of Lidice, over 15 years of age, were shot – the place of execution is marked by a simple cross – the women and children were separated from each other and taken off to concentration camps; every building in the town was burnt down and the whole area flattened.

Immediately following the end of the Second World War, work began on building a new Lidice which today is a housing development of uniform design close to the original town, with lawns laid out in between the houses. There is a room in the national monument's museum which is dedicated to the victims of the Nazis and roses from all over the world were planted in the garden of friendship and peace. The sculpture of "Woman with Child" by B. Stephan marks the site of the mass grave. Where the vicarage once stood, there now stands a memorial by K. Lidický.

Location
20 km (13 miles) NW of Prague

Lobkowitz Palace (Lobkovický palác)   E2

The Lobkowitz Palace, restored in the mid-1970s, is now occupied by the German Embassy. Built at the beginning of the 18th c. by Giovanni Battista Alliprand in Early-Baroque style, it was altered and had an additional storey added to the side-wings by I. Palliardi in 1769. In 1753 the palace was bought by the Lobkowitz family, whose arms appear on the gables.

The main façade is finely articulated, with a massive doorway, a pediment decorated with sculpture and an attic storey with statues. A ceiling fresco in the stair-well depicting the victory of good over evil is said to be the work of J. J. Steinfels, who was also probably also responsible for the richly painted decoration in the interior (c. 1720).

Still more interesting is the rear façade, with a cylindrical projection and a *sala terrena* leading into the grand courtyard. On the roof above the projecting bay is an unusual architectural feature – an ornamental pond. From the grand courtyard, which is enclosed by the three wings of the palace, a gateway enriched by sculpture (abduction of Proserpina and Orechteia) gives access to the English-style park (originally laid out by J. J. Kapula at the time the palace was built and re-styled at the end of the 18th c.)

Location
Vlašská 19,
Malá Strana Praha 1

Trams
12, 22

Loreto (Loreta)   D2

From Hradčany Square (see entry) a street lined with old burghers' houses, Loreto Street (Loretánská ulička) – in Loreto Square, which Peter Parler, the architect of St Vitus's Cathedral, Hradčany, once lived – leads to Loreto Square, which ranks with Knights of the Cross square (see entry) and one or two other as one of the most beautiful features of Prague.

Along the south-west side of the square extends the 150 m (490 ft) long façade, articulated by 30 tall pilasters, of the Cernin Palace (1669–97; see entry).

On the east side of the square, which falls steeply towards the north, is the old shrine and pilgrimage centre of Loreto.

Location
Loretánské náměstí Hradčany
Praha 1

Trams
22, 23

Prague's Loreto Shrine is the best known in Bohemia. The Loreto

Loreto Shrine

Loreto

Loreto Shrine

cult, a branch of mariolatry, came from Italy to the countries of eastern Europe in the mid 15th c. Its origin was based on the Bible story of the Holy Family's home in Nazareth, where the Archangel Gabriel appeared before the Virgin and told her of the birth of Jesus. According to a 13th-c. legend, the house (Casa Santa) was transported to Italy to afford protection against non-believers, and the story was widely spread among Catholics in the Baroque period. During the Counter-Reformation some fifty Loreto shrines were built in Bohemia on the model of the Santa Casa shrine at Loreto in Italy, in order to increase the appeal of Catholicism to ordinary people.

History of the building

The main front (1721 onwards) was designed by Christoph and Kilian Ignaz Dientzenhofer. The bell-tower, in Early-Baroque style, is older; it contains a carillon (installed by P. Neumann in 1694) which plays Marian hymns every hour in summer; the 27 bells, with a total weight of approximately 1540 kg (3380 lbs) were cast in Amsterdam in 1694 by Claude Fremy for the wealthy merchant Eberhard von Glauchau.

Loreto Chapel

In the two-storeyed cloister with its central fountain is the Loreto Chapel (Loretánská kaple), the architectural and spiritual heart of the whole complex. The chapel was founded by Countess Benigna Katherina von Lobkowitz in 1626. It was built by Giovanni Battista Orsi from Como who completed the work in 1631. The sculpture and stucco reliefs, based on Italian designs depicting the Lives of Mary, Old Testament prophets and pagan sibyls, are by G. Agosta, G. B. Colombo and G. B. Cometa (1664). The legend of the shrine is depicted on the east wall of the chapel.

Casa Santa

The Casa Santa contains pictures of scenes from the life of the Virgin Mary which were painted in 1695 by František Kunz, the

114

Loreto Shrine
Loreta

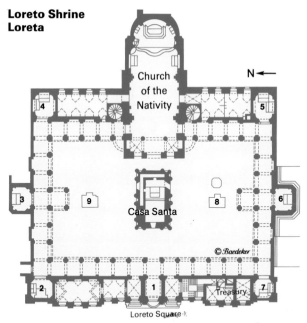

N ←

Church of the Nativity

4

5

3

9

Casa Santa

8

6

© Baedeker

2

1

Treasury

7

Loreto Square

1 Main entrance
2 St Anne's Chapel
3 Chapel of St Francis
4 Chapel of St Joseph
5 Holy Cross Chapel

6 Chapel of St Anthony of Padua
7 Chapel of the Sorrows of Mary
8 Fountain of the Assumption
9 Fountain of the Ressurection

artist from the Lesser Quarter, a silver altar and a carved cedar-wood Madonna. It is framed by a silver garland of two or three groups of winged cherubs and is attributed to the Prague gold-smith Markus Hrbek. The toal weight of the silver decoration in the Casa Santa amounts to over 50 kg (110 lbs).

The two fountains, known as "Mary's Ascension into Heaven" (1739; copy by V. Suchardaa) and the "Ressurection of Our Lord" (1740) by J. M. Brŭderle, stand in the courtyard on either side of the Casa Santa.

Fountains

The lower storey of the cloister which was begun in 1634, has frescoes by F. A. Scheffker (1750; restored in 1882). Along the walls of the arcades are several altars with figures of saints by unknown artists in the Baroque style which was beginning to lose its popularity.

Cloister

The cloisters are bordered by seven chapels:
The ceiling fresco in St Anne's Chapel by F. A. Scheffler (1750) is of the Virgin Mary in the Temple. The St Francis Seraph Chapel built in 1717 by Christoph Dientzenhofer has a high altar by Matthias W. Jäckel with a picture of the saint from the workshops of the great Baroque artist, Johann Peter Brandl. The high altar in the chapel of St Joseph, founded in 1691, is adorned with a picture of the Holy Family and a vault fresco by an unknown

Chapels

Casa Santa

Baroque artist depicts the 12-year old Jesus in the Temple. The dome fresco ("St Dismas and the Virgin Mary" – 1750) in the Holy Cross Chapel is a work by F. A. Scheffler. The builder, Cristoph Dientzenhofer and sculptor, M. W. Jäckel worked together on the St Anthony of Padua Chapel (1710–12). Sebastian Zeiler did the paintings for the altar which was made by M. W. Jäckel. The pietà on the high altar of the Mater Dolorosa Chapel dates back to the Hussite Wars, the wall-paintings, however, are the work of F. A. Scheffler.

Church of the Nativity

On the east side of the cloister is the Church of the Nativity (Kostel Narozeni Páně), begun by Christoph Dientzenhofer in 1717, continued by his son Kilian Ignaz and completed by Georg Aichbauer in 1735. The principal feature of the bright interior is the High Altar, with an altarpiece ("The Birth of Christ") by Johann Georg Heintsch. The ceiling painting in delicate colours is of "Christ in the Temple" by Wenzel Lorenz Rainer shows the influence of Venetian Illusionism. Ceiling frescoes ("Adoration of the Shepherds" and "Adoration of the Kings") are by J. A. Schöpf, the principal of the Prague artists guild.

Treasury

The Treasury, which was moved to the west wing in 1962 and restored in 1984, is on the upper floor of the cloister. Apart from vestments and liturgical items it also contains valuable 16th–18th c. monstrances, including the tiny Pearl Monstrance which is decorated with 266 diamonds and a ruby as well as pearls (1680) and the silver-gilt Ring Monstrance, made in 1748 and decorated with 492 diamonds, 186 rubies, a sapphire and 24 pearls as well as emeralds, amethysts and garnets. On display also is the famous Diamond Monstrance (with more than 6200 diamonds), inherited from Ludmilla Eva Franziska Kolowrats and made in

Vienna in 1699 by the court jewellers Matthias Stegner and
Johann Kánischbauer.

Capuchin Friary (Kapucínský klášter)

On the north side of Loreto Square, down the hill from the Černín
Palace, is the first Capuchin Friary to be built in Bohemia (1600–
02), in the plain architectural style of Capuchin houses. An
enclosed corridor at first-floor level links it with the Loreto Shrine
opposite.
Adjoining the friary is a simple church dedicated to the Virgin,
which originally had 14 Gothic panel-paintings of unknown origin (now in the National Gallery). At Christmas the church attracts
large numbers of worshippers with its beautiful Baroque crib
(Nativity scene).

From the Loreto Shrine a street runs north to the "New World"
(Novy svet), a picturesque old corner of the town on Hradčany
Hill. More and more artists and students are making it their home.
The House of the Golden Horn was occupied by the great astro-
nomer Johannes Kepler in 1600. The Baroque House of the Gold-
en Pear is now an attractive restaurant.

"New World"

Loreto Street (Loretánská ulička)   E1/2

This street runs through the Lesser Quarter in an east-west direction between Hradčany Square (passing in front of the castle) and
the Loreto. House No. 1 is of historical interest as it was the Town
Hall of the Hradčany up to 1784. The house was built at the
beginning of the 17th c. by K. Oemichen of Oberheim, after the
Hradčany Quarter had been elevated to a Royal Town in 1598.
The remains of Imperial coats-of-arms of the Hradčany above the
doorway reflect this glowing period in the town's history.

Location
Praha 1
Hradčany
Loretánská ulička

Trams
22, 23

The Hrzán Palace (No. 9) also deserves special mention as it was
originally a Gothic house owned by the famous master-builder
Peter Parler. The building was reconstructed in Renaissance style
in the mid 16th c. and at the end of the 18th c. received its
Late-Baroque façade. At the beginning of the 20th c. the famous
art school, run by Ferdinand Engelmüller (1867–1924) was
housed here. The bust in the courtyard is in his memory. In the
mid 1950s the palace was re-built for the final time and it is now
used for state administrative purposes.

Hrzán Palace

Maisl Synagogue

See Josefov

Malá Strana

See Lesser Quarter

Mánes Exhibition Hall   F4

The Mánes Exhibition Hall, in Constructivist style, on the site of
the old Sitkov mill, was built by O. Novotný in 1930 for the group
of artists associated with Josef Mánes (see Notable Personal-

Location
Gottwaldovo nábřeží 20,
Nové Město, Praha 1

117

Mánes Exhibition Hall

Metro
Anděl

Trams
4, 6, 7, 9, 16

ities). It was carefully restored between 1980 and 1987 and is now the home of the Artists' Union. Attached to the exhibition hall are a restaurant and a café with a garden terrace.

Adjoining the Mánes Exhibition Hall is the Šitek Water-Tower (see entry).

Martinitz Palace

See Hradčany Square

Mělník

Location
38 km (24 miles) north of Prague

Lying amid vineyards on the right bank of the river, the country town of Mělník (222 m, 729 ft above sea level; pop. 14,000) looks over the confluence of the Vltava and the Elbe. It is the wine-producing centre of Bohemia and each year on the last week-end in September a huge wine-festival is held in the town. Mělník's own well-known "Ludmilla wine" is named after the country's first martyr and patron saint. During his reign, Charles IV imported the Burgundy vine and cultivated it in the vineyards above the Vltava.

The Gothic diocesan church of St Peter and Paul (15th c.) and an enormous castle built in the 14th c. dominate the skyline of the town which has enjoyed civic rights since 1274. It is possible for over 6,000 hl. of wine to be stored in the wine cellars of the former Lobkowitz Castle which was converted into a Renaissance palace during the 16th c. It now houses a collection of valuable paintings from the Baroque period as well as the Regional Museum with local history and exhibitions of viticulture. (Tues.–Sun., May–Aug. 8 a.m.–5 p.m: Sept., Oct., Apr. 9 a.m.–4 p.m.). The old 14th c. Town Hall (re-built in Baroque style 1756–93) and the remains of the town's fortifications dating back to the 13th c., with the "Prague Gate" (*c.* 1500) are worth seeing.

Liběchov Castle

The town of Liběchov (Liboch, 163 m/533 ft above sea-level, pop. 1300) lies surrounded by vineyards roughly 7 km/4 miles north-west of Mělník on the right bank of the Elbe. It is the gateway to the romantic area of the Liboch Estates. Its castle built in Renaissance style in the 16th c. now houses an art gallery (Apr.–Sept. Tues.–Sun.). There is a 19th c. statue carved from outcrops of rock in the Baroque garden wing (18th c.)

Kokořín Castle

This castle which is a splendid example of the romantic Neo-Gothic style so popular at the beginning of this century, towers amid dense woodland above the Psovla just under 17 km/10 miles north-east of Mělník. The river passes between rugged limestone rocks as it makes its way through the charming countryside of the Kokořín Valley. Originally built at the beginning of the 14th c., the castle was severely damaged during the Hussite Wars and restored between 1911 and 1918.

Legend has it that the castle is haunted by the Black Knight who had his elderly house-maid, Hate, hurled into the castle moat in order that a young lady of noble birth might be freed from the power of the cruel castellan. Since that time the Black Knight has to hunt through the hours of darkness until the world is rid of human greed.

Mladá Boleslav (Jungbunzlau)

This country town (230 m/755 ft above sea-level, pop. 30,000) is best known for the motor industry. Skoda cars were originally made in a bicycle workshop here and the very first motorcycles went into production in 1899. The manufacture of cars began six years later but the brand name was only introduced in 1925 when the original firm of Laurin and Klement was set up in the Skoda works. An upward swing in production was noticed particularly after 1964 when output from the new works (80 ha.) rose. A fascinating exhibition about the development of the motor car may be found in the historic rooms of the 16th c. castle. Notable features of the Old Town, which stands on a rocky hill overlooking the Iser (Jizera) are the Old Town Hall, dating from 1559 and the church (built originally in the 15th c. and renovated in Baroque style at the beginning of the 18th c.) in the Old Town Square. The town's museum is in the former Protestant Church of the Bohemian Brotherhood ("Temple" mid 16th c.). The Czech avant-garde architect J. Kroha (d. 1974) designed the housing estate which was put up here in 1923 establishing a completely new style in house-building.

Location
56 km (35 miles) north-east of Prague

Morzin Palace (Morzinský palác) E2

Location
Nerudova ulice 5,
Malá Strana, Praha 1

Metro
Malostranská

Trams 12, 22

In the Morzin Palace (1714), one of the finest Baroque palaces in the Lesser Quarter (see entry), the Baroque architecture of Giovanni Santini combines with the sculpture of Ferdinand Maximilian Brokoff to form a harmonious whole. The balcony is supported on figures of Moors – heraldic emblems of the Morzin family. Above the doorway are allegories of Day and Night. The building is now occupied by the Rumanian Embassy.

Mozart Museum

See Bertramka

Municipal Museum (Muzeum hlavního města Prahy) D6

Location
Sady Jana Švermy,
Praha 8, Karlín

Metro
Florenc

Trams
3, 5, 8, 10, 13, 19

Opening times
Tues.–Sun. 9 a.m.–noon
1 p.m.–5 p.m.

The Municipal Museum was built in 1898 by Antonín Balšánek and A. Wiehl in neo-Renaissance style. The decoration on the façade is by a number of different sculptors.
In the hall by the staircase hangs the face painted by Josef Mánes for the astronomical clock of the Old Town Hall (see entry), with the signs of the zodiac on the inner ring and the twelve months, symbolised by scenes of country life, on the outer ring.
The museum, originally founded in 1884, illustrates the economic, architectural and cultural history of Prague down the centuries, with furnished rooms, historic costumes, jewellery, pottery and sculpture from the Prague area and a collection of Prague house signs (see entry). An item of particular interest is a model of the town, measuring 20 sq. m (215 sq. ft), by the lithographer A. Langweil (1830), which depicts in minute detail the houses, churches and palaces of early 19th c. Prague.

*Museum of Applied Art (Umělecko-průmyslové muzeum) D4

Location
Ulice 17. listopadu 2,
Staré Město, Praha 1

Metro
Staroměstská

Buses
133, 207

Tram
17, 18

Opening times
Tues.–Sun. 10 a.m.–6 p.m.

The Museum of Applied Art was founded in 1884. The present neo-Renaissance building, on the west side of the Old Jewish Cemetery (see Josefov), was erected in 1897–1901 (architect Josef Schulz).
The museum has a world-famed collection of glass, porcelain and pottery ranging from ancient times to the present day, furniture of the 16th–19th c. and goldsmith's work of the 15th–19th c. Other fields represented are textiles, measuring instruments, bookbinding, commercial art, small bronzes and coins (illustrating their historical development, with examples dating back to 700). From time to time there are well-prepared special exhibitions.
The museum has a specialised library (open to the public) on art history and applied art, including a collection of 15th c. parchments.

Museum of Military History

See Hradčany Square, Schwarzenberg Palace

Na Příkopě (On the Moat)

The street named Na Příkopě, known to German-speakers as the Graben, is the busiest street in Prague, running between the lower end of Wenceslas Square (see entry) and Revolution Square (Náměstí revoluce). Na Příkopě, Wenceslas Square, Národní třída and the side streets opening off them are the commercial centre of Prague, known as the "Golden Cross", with banks, office blocks, shops, cafés, etc. The street follows the line of a stream which later became a moat between the Old and New Towns and was filled in 1760.

Location
Praha 1

Metro
Můstek

Coming from Wenceslas Square, the Old Town lies on the left, the New Town on the right. The Head Office of the ČKD Prague engineering works, opened in 1983, is on the corner of Na Příkopě and Na můstku. At the top of this striking, modern steel-girder construction is a huge clock taken from the building which previously occupied the site. There are several restaurants here, open to the public, including a coffee-house on the 5th floor from where there is a pleasing view over the Old Town.

ČKD Prague

Immediately opposite is the oldest department store in Prague, the textile house Dům elegance (No. 4). Built at the end of the 19th c. to the design of the Viennese architect T. Hansen, it has a stone façade in Italian Late-Renaissance style.

Textile House
Dům elegance

Just a short distance away is the Sylva-Taroucca Palace (No. 10). This gem of Bohemian Late-Baroque architecture is now occupied by function rooms and a restaurant (tables in the garden in summer).

Sylva-Taroucca Palace

The palace, built for Prince Ottavio Piccolomini by Anselmo Luragno in 1743–51 to the design of Kilian Ignaz Dientzenhofer, shows the influence of French Classical architecture, with two courtyards, a carriage entrance flanked by columns, a garden and a riding-school. The decoration of the richly articulated façade, with sculptures from Greek mythology and splendid vases, and the Rococo staircase was the work of Ignaz Platzer the Elder; the stucco-work in the interior is by Carlo Bassis. The frescoes on the stairway (including allegorical pictures of the four seasons) are by V. B. Ambrozzi.

The neighbouring neo-Romantic "House of the Black Rose" (U černé růže) once belonged to the University of Prague. From 1411 onwards German supporters of the great Reformer Johann Hus met here, and followers of the Black Rose (including J. Craendorf and P. Turnow) played an important role in the spread of the Hussite Movement in Germany.

House of the Black Rose

The Children's Store was designed in the Constructivist style at the end of the 1930s by L. Kysela; in 1950 the building was completely altered.

Children's Store
Dětský Dům

Holy Cross Church, the only Empire church in Prague, was built by J. Fischer in 1816–24 for the Piarist Teaching Order. The old Piarist school is in Panská ulice.

Holy Cross Church
(Kostel svatého Kříže)

At No. 18 is the office of the Čedok travel agency. It occupies the old Land Bank (Zemská banka) built in 1912.

Travel Agency
(Čedok)

House No. 20, which retains its Bohemian Renaissance style, has allegorical mosaics designed by M. Aleš and reliefs from the workshops of C. Klouček and S. Sucharda. The Prague Information and Document Service (Praška informační služba) is able to tell you all you need to know about the Czech capital.

Prague Information Service

Na Příkopě in the Golden Cross

Čedok headquarters

Slav House (Slovanský dům)	The Baroque Přichovských Palace, built 1695–1700 by Count Jean B. Vernier de Rougemont was the "German House" between 1875 and 1945 and served as the meeting place of Prague's German population. After the Second World War, it was renamed "Slav House" (Slovanský dům) and is now used for cultural events. It received its Classical appearance during modifications by the Tuskany family towards the end of the 19th c.
Palace of the State Bank	No. 24 houses the Palace of the State Bank of Czechoslovakia, built in 1938. It stands on the site of the two famous hotels, the "Blue Star" and the "Black Horse" where Europe's leading figures (among them Franz Liszt and Frederick Chopin) stayed in the 19th c. The peace treaty between Austria and Prussia was signed in the Blue Star Hotel in 1866.
Moscow House	The Moskva (Moscow) Restaurant – offering good Russian cuisine – and the Arbat Tavern are in No. 29, Moskva House, which was remodelled in 1960 to the design of E. Linhart.

National Memorial on St Vitus's Hill   E7

(Národní památník na hoře Vitkově)

Location
Na vrchu Žižkově, Žižkov, Praha 3

Buses
133, 168

The National Memorial on St Vitus's Hill, in the eastern district of Žižkov, is best reached by way of Třída Vitězného února, Husitská třída and the street called U Památníku (At the Monument), and thereafter by steps.

The National Memorial was built in 1927–32, but the interior was completed only in 1948 and subsequent years. Above a terrace which affords extensive views of Prague and the hills to the west rears a monumental equestrian statue by Bohumil Kafka (1950) of the Hussite General Jan Žižka, who defeated the Imperial forces here. Beyond it is a massive mausoleum for the Presidents of the Republic and other politicians of the post-war Communist period. The upper part is a Mourning Hall (with a large organ), and below this is the mausoleum proper, with marble sarcophagi (containing only urns) and a columbarium. Here, too, is a hall commemorating those who fell in the First World War. Under the equestrian statue lies the Tomb of the Unknown Soldier of the Second World War (one who fought at the Dukla Pass: see Town Hall of Old Town).

At the foot of St Vitus's Hill, at the near end of U Památniku (on the right), is the Military Museum of the Czechoslovak Army (Vojenské muzeum Československé armády). Weaponry, uniforms, battle plans and paintings depicting the Second World War are on display here.
The Museum is open between May and October from 9.30 a.m. to 4.30 p.m., Tuesday to Sunday.

Opening times
Apr.–Oct., 9 a.m.–5 p.m.;
Nov.–Mar. 10 a.m.–4 p.m.;
Museum: Tues.–Sun.
9.30 a.m.–4.30 p.m.

National Museum (Národní muzeum) D6

The National Museum, founded in 1818, is Czechoslovakia's oldest museum, among its founders were famous people, such as Count Kašpar Sternbeck, Josef Dobrovský and Fratišek Palacký. It incorporates the Ethnographic Museum (see entry), the Music Museum, the Náprstek Museum with collections from Asian, African and American cultures, and the Museum of Physical Culture and Sport, all housed in separate buildings.

Location
Václavské náměstí (Wenceslas Square), Nové Město, Praha 1

Metro
Muzeum

The National Museum in Wencelas Square

National Museum

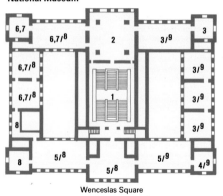

Wenceslas Square

National Museum
Národní muzeum

FIRST FLOOR (PRVNÍ POSCHODÍ)

1 Statue of King George of Poděbrad (1420–71), by Ludwig Schwanthaler
2 Busts of men who have made valuable contributions to science or to the museum
3 Prehistory and archaeology
4 Numismatic department
5 Special exhibitions
6, 7 Mineralogy and petrography

SECOND FLOOR (DRUHÝ POSCHODÍ)

8 Zoology
9 Palaeontology

■ Statues

■ Busts on bases

● Busts on the walls

1 Beneš z Loun (1454–1534), architect
2 Ferdinand Maximilian Brokoff (1688–1731), sculptor
3 Miroslav Tyrš (1832–84), art historian
4 Karel Škréta (1610–74), painter
5 Josef Mánes (1820–71), painter
6 Jaroslav Heyrovský (1890–1966), founder of Polar exploration
7 František Palacký (1798–1876), historian and politician
8 Tomáš Garrigue Masaryk (1850–1937), President of Czechoslovak Republic
9 Karel Havlíček Borovský (1821–56), journalist
10 Ctibor Tovačovsky z Cimburka (1438–94), politician
11 František Josef Gerstner (1758–1832), physicist and technician
12 Bohuslav Balbín (1621–88)
13 Jan E. Purkyně (1787–1869)
14 Josef Ressel (1793–1857), inventor of the screw-propeller
15 Josef Dobrovský (1753–1829)
16 Pavel Josef Šafařík (1795–1861)
17 František Martin Pelci (1734–1801), historian
18 Josef Jungmann (1773–1847)
19 Karel ze Žerotina (1564–1636)
20 Viktorin Kornel ze Všehrd (1460–1520), jurist
21 František Škroup (1801–62), composer of the National Anthem
22 Jan Hus (1371–1415), Reformer
23 Antonín Dvořák (1841–1904), composer
24 Bedřich Smetana (1824–84), composer

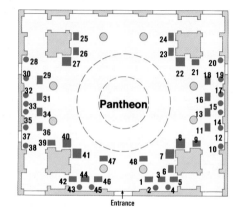

Entrance

25 Stanislav Kostka Neumann (1875–1947), poet
26 Julius Fučík (1903–43)
27 Jan Amos Komenský (1592–1670), pedagogue and Reformer
28 Daniel Adam z Veleslavína (1546–99), historian
29 Pavol Hviezdoslav (1849–1921), poet
30 Karel Jaromir Erben (1811–70), poet and historian
31 Jaroslav Vrchlický (1853–1912), poet
32 Antonín Jaroslav Puchmajer (1769–1820), poet
33 Jan Kollár (1793–1852), poet
34 Svatopluk Čech (1846–1908), poet and writer
35 Václav Matěj Kramerius (1753–1808), writer
36 Alois Jirásek (1851–1930), writer

37 František Ladislav Čelakovský (1799–1852), poet
38 Tomáš ze Štítného (1333–1405), religious philosopher
39 Petr Bezruč (1867–1958), poet
40 Jan Neruda (1834–91), poet
41 Kašpar Šternberk (1761–1838), co-founder of National Museum
42 Václav Hollar (1607–77), engraver
43 Václac Vavřinec Reiner (1689–1743), painter
44 Josef Václav Myslbek (1848–1922), sculptor
45 Petr Jan Brandl (1668–1735), painter
46 Mikoláš Aleš (1852–1913), painter
47 Eliška Krásnohorská (1847–1926), poetess
48 Božena Němcová (1820–62), writer

The main building, with its recently regilded dome, stands at the upper end of Wenceslas Square (see entry). In neo-Renaissance style, it was built between 1885 and 1890 by Josef Schulz, and houses the Natural History and Historical Museums and the Library of the National Museum (over 1·3 million volumes).
In the Pantheon, a domed hall two storeys high, are statues and busts of distinguished Czechs. The mineralogical, botanical and zoological collections are in the side wings. Of particular interest in the department of history and archaeology is a large collection of coins and medals as well as exhibits illustrating the history of the Czech theatre and puppet theatre.

Tram
11

Opening times
Sat., Sun., Wed. and Thurs.
9 a.m.–5 p.m., Mon. and Fri.
9 a.m.–4 p.m.

National Museum of Technology (Národní technické muzeum)   C5

The National Museum of Technology, on the northern slopes of Letná Hill (see entry), offers a wide-ranging survey of the development of cinematography in more than 50 countries, radio and television, transport and mining. A particularly impressive feature is a 600 m (650 yd) long mine shaft. In the main hall are aircraft and locomotives, as well as two cars belonging to the Emperor Francis Joseph in which the heir to the throne, Francis Ferdinand, drove to Sarajevo, where he was assassinated in 1914. In the courtyard to the right of the museum there are also a number of aircraft.

Location
Kostelní ulice, Holešovice,
Praha 7

Tram
1, 8, 25, 26

Opening times
Tues.–Sun. 10 a.m.–5 p.m.

National Theatre (Národní divadlo)   F4

The neo-Renaissance National Theatre, built by Josef Zítek in 1868–81 was burnt down shortly after the first performance in the theatre, but was rebuilt by Josef Schulz within two years, the cost being met by contributions from the public. On 18 November 1883 the first season began with the splendid première of B. Smetana's opera "Libussa". The theatre "incarnates all the yearnings and aspirations of a people which after long slumber was returning, full of energy and enthusiasm, into the European intellectual community" (V. Volavka).
The theatre, restored between 1976 and 1983, is generously adorned with sculptures, and free-standing statues serve to complete the silhouette of the massive building. The figures on the attic loggias (the goddesses of victory; Apollo and the muses) are by Bohuslav Schnirch and the allegories depicting opera and drama on the western side as well as the statues of Záboj and Lumír, in the niches on the north façade, are by the Viennese sculptor Anton Wagner. The allegorical paintings of musical comedy and drama above the side entrance are works by the artist, Josef Václav Myslbek. The sculptor later made the busts of famous people for the portrait gallery and the allegorical painting of music for the large foyer (1913).
All the leading artists of the day contributed to the interior decoration of the theatre. The frescoes in the auditorium, loggia and president's suite portray mystical and historical themes or the world of the theatre. The ceiling fresco in the auditorium has eight allegories depicting the art of František Ženíšek whilst the lunettes in the loggia, looking over National Street, were painted by Josef Tulka. The large foyer on the first balcony has an impressive painting by F. Ženíšek of the "Golden Age, Decline and Revival of Art". Together with Mikoláš Aleš, the artist also painted the cycle in 14 tympanums entitled "The Homeland" as well as the four wall-paintings "Pagan Myths", "History", "Life" and

Location
Národní třída (corner of
Smetana Embankment),
Nové Město, Praha 1

Trams
9, 17, 18, 21, 22

National Museum of Technology

"Folk Music". The lunettes in the connecting passage of the large foyer were decorated by Adolf Liebscher. In the President's wing, works may be seen by Vojtěch Hynais ("Four Seasons"), who was also responsible for the stage curtain (allegorical portrayal of the reconstruction of the theatre); by Julius Mařák and by Václav Brožík (themes from Czech history).

"New Scene"

The modern structural complex known as the "New Scene" was erected in 1983 to complement the historical building.

Neruda Street (Nerudova ulice)   E2

Location
Praha 1, Malá Strana,
Nerudova ulice

Metro
Malostranská

Trams
12, 20

This street which was, in the past, the main approach road begins in the Lesser Quarter Square and climbs steeply to the Hradčany. The old Spur Road with, in the main, townsmen's houses in Late-Baroque style is one of the finest in Prague. It formed the background for the "Lesser Quarter Stories" written during the last century by the Czech author, Jan Neruda, after whom the street is named. The writer of unimaginative short stories lived in the house of the Two Suns (U dvou slunců, No. 47) between 1845 and 1857 and a commemorative plaque is on the Early Baroque façade. The Baroque Morzin Palace (No. 5) is now the Romanian Embassy. The front of the house on the opposite side is decorated with a Baroque sign featuring a red eagle supported by two angels. Three crossed violins make up the attractive sign on house No. 12 where the Prague family of violin-makers the Edlingers, used to live. It is now a wine-bar. Originally Gothic, Valkounsch House, which was later remodelled in Renaissance

The National Theatre▶

"The three little fiddles" in Neruda Street

style, has the artist, J. B. Santini-Aichl to thank for its Late-Baroque appearance. Interesting house-signs may be seen on the Renaissance house, "The Golden Cup" (U zlaté číše; No. 16) and on the Baroque house of "St John of Nepomuk" (U svatého Jana Nepomuckého; No. 18). The Baroque Thun-Hohenstein Palace (No. 20) now houses the Italian Embassy. The so-called Kajetan dramas were performed in Czech in the refectory of the old Theatine or Kajetan Convent (No. 24) during the first half of the 19th c. Neruda's story, "A week in a silent house" is set in the house next door, "To the Rocking Donkey" (Osel u koébky). The old Lesser Quarter chemist's shop (restored in 1980) in the house of The Golden Lions (U zlatého lva; No. 32) is now a Pharmaceutical Museum displaying items from all over Bohemia. The Baroque Bretfeld Palace (Bretfeldský palác; No. 33), where such distinguished guests as W. A. Mozart and G. Casanova once stayed, is known as "Summer and Winter". The first Lesser Quarter pharmacy was in Golden Horseshoe House (U zlaté podkovy; No. 34), the door of which is guarded by St Wenceslas. The Baroque burgher palace (U bilé labutě; No. 49) has an ornate white swan as its house-sign. From here it is just a short distance via the Ke Hradu steps to Prague castle.

New Town Hall

See Charles Square

Nostitz Palace (Nostický palác) E3

The Nostitz Palace, a quadrangular structure built round a courtyard, occupies the south side of Maltese Square in the Lesser Quarter (see entry). It now houses the Netherlands Embassy and various departments of the Ministry of Culture.

Location
Maltézské náměstí 471,
Malá Strana, Praha 1

This Baroque palace was built for Johann Hertwig von Nostitz, probably by Francesco Caratti, in 1658–60. The dormer windows and the statues of emperors (copies) (from the workshop of F. M. Brokoff) were added in 1720, the columned doorway (by Anton Hafenecker) after 1765. Notable features of the interior are the beautiful courtyard and the ceiling-frescoes with mythological motifs in the principal apartments (by W. B. Ambrozzi, *c.* 1757). The palace also houses the Dobrovský Library (formerly the Nostitz Library with over 15,000 volumes).

Trams
12, 22

Old Jewish Cemetery

See Josefov

Old-New Synagogue

See Josefov

Old Town (Staré město)

D/E4/5

See plan
pp.130/131

The historic Old Town Square with the massive memorial to Jan Hus, the great Bohemian reformer, is the hub of Prague's Old Town. By the end of 1987, the enormous square and all its old buildings had been completely restored. Houses with exquisite, brightly coloured façades grace the north and south sides whilst the east is dominated by the Gothic Týn Church (The old Church of the Prague Utraquists) and the Late Baroque Kinsky Palace. St Nicholas's Church in the Old Town, a splendid example of Dientzenhofer Baroque, can be seen on the north-western side of the square. Standing in the south-west corner is the Town Hall of the Old Town. Crowds of people gather in front of its tower on the stroke of the hour to admire the procession of figures on the famous Astronomical Clock.

Going eastwards, the pedestrian zone in the Celetna, which has also been lovingly restored, joins up with the Powder Tower. The original building on the site was one of the 13 gateways which formed part of the fortifications of Prague's Old Town. Immediately next to it is the splendid House of Representation built at the turn of the century in Sezession style, together with the Kotva department store which is always crowded. Interesting exhibitions are staged in Hibernian House on the opposite side. The bustling Na Příkopě (On the moat), where the Czech Travel Agency, Čedok, has its headquarters, leads on from here to Wenceslas Square.

Tiny little streets, so emphatically described in the novels of Franz Kafka, head off southwards from the Old Town Square to Charles University, the oldest in Europe. A few steps further on is the Týl Theatre where Mozart's "Don Giovanni" was performed for the first time over 200 years ago. The continuation of the "Royal Route" from the Old Town Square to Prague Castle has also been made into a pedestrian zone. Beginning in Charles Street (Karlova), it proceeds to the Old Jesuit College in the Clementinum and then to Knights of the Cross Square. The route then passes over Charles Bridge with its sixteen arches – the oldest bridge on

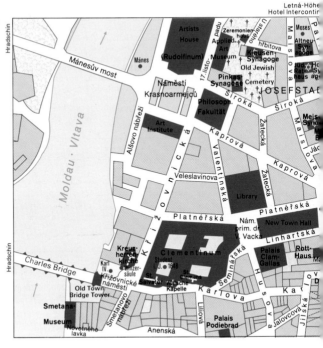

the Vltava – and on to the Lesser Quarter. Pariska Street which was designed in Art-nouveau style runs from the north end of the Old Town Square to the Josefov, the old Jewish Quarter of Prague named after the Emperor Joseph II. It was demolished at the end of the 19th c. and all that remains today are the Town Hall, six synagogues and the old Jewish cemetery which is now the Jewish State Museum.

The internationally renowned music festival known as "Prague Spring" takes place each year in the Artists' House, a short distance to the west on the banks of the Vltava. .The Lesser Quarter, on the other side of Mánes Bridge, stretches as far as the massive castle complex of the Hradčany.

Old Town Bridge Tower

See Charles Bridge

Old Town Square (Staroměstske náměstí)   E4

Location
Staré Město, Praha 1

Metro Staroměstská

Old Town Square ranks with the Hradčany (see entry) as one of the two most historic places in Prague. This spacious square (9000 sq. m (11,000 sq. yd)) was the market-place of the Old Town in the 11th and 12th c., and lay on the route of the traditional

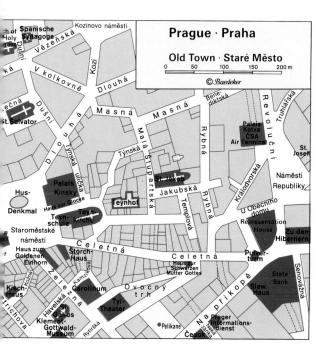

coronation procession of the Bohemian kings from the Vyšehrad (see entry) to the Hradčany.

The square has been the scene of great events, both glorious and tragic.

1422: The radical Hussite preacher Jan Želivský, who with his supporters from the poor quarters of the city had stormed the New Town Hall in 1419 and, with the First Defenestration of Prague, had given the signal for the Hussite Wars, is executed (memorial tablet and bust on east side of Town Hall).

1437: The Hussite officer Jan Roháč of Dubá and 56 other Hussites are executed.

1458: George of Poděbrad is elected as King of Bohemia in the Town Hall (picture by V. Brožík in Town Hall). His reign was the heyday of Bohemian independence.

1621: The 27 leaders of the Protestant rising of the nobility are executed on the orders of Ferdinand II (commemorative tablet on east side of Town Hall).

1915: 500th anniversary of the death of Jan Hus. The Jan Hus Monument (by Ladislav Šaloun), with the inscription "The truth will prevail", is unveiled.

1918: End of the monarchy. The people of Prague demonstrate in the square, calling for a Socialist Czechoslovakia.

1945: End of the Second World War; the Soviet army is given an enthusiastic reception by large crowds.

1948: Speech by Klement Gottwald on the balcony of the Kinsky Palace (see entry) proclaiming the Communist accession to power.

1968: End of the "Prague Spring". The tanks of the Warsaw Pact forces are received with Molotov cocktails and the Jan Hus Monument is draped in black.

In the basements of the buildings in Old Town Square (see Town Hall of Old Town, St Nicholas's Church, Kinsky Palace, Týn Church, St James's Church) are many remains of Romanesque houses. After the devastation caused by the flooding of the Vltava in the 11th and 12th c. the level of the Old Town was raised by the deposit of additional soil and the new houses were built on top of the old ones.

1988: Thousands protest on the 20th anniversary of "Prague Spring". There is a huge demonstration through the Old Town against the occupation by Warsaw Pact forces and the suppression of the Reform Movement. Demands are made for freedom and civil rights.

The Old Town Square together with its magnificent buildings including the Town Hall of the Old Town (see entry); the Týn Church (see entry; still undergoing alteration) and the Týn School, the Kinsky Palace (see entry) and the Church of St Nicholas in the Old Town (see entry) were all fully restored by the end of 1987.

The newly constructed façades on the buildings around the square are now finished in pastel tones. The tall house, "To the Stone Bell" (U kamenneho zvonu, No. 13) stands next to the Kinsky Palace. It was reconstructed in its present Gothic style prior to 1987 and was probably converted into a palace during the mid 14th c. by the wife of John of Luxembourg. Alterations in Baroque style followed during the 17th and 18th c. Three of the

Jan Hus Memorial in the Old Town Square

Štorch House

A Baroque façade

original statues which pay tribute to the monarchy and its supporters have survived (copies have been made).

"Stork House" (No. 6) in neo-Renaissance style, stands at the end of the row of splendid Baroque houses on the south side of the square. A 14th/15th c. Gothic house stood on the site originally but this was replaced in 1897 by the present building, designed by B. Ohmann. The painting on the front of the house of St Wenceslas on horseback is the work of M. Aleš. The house "At the sign of the Golden Unicorn" (U zlatého jednorožce; No. 20) outgrew its Romanesque core in the 14th c. and was remodelled in Late Gothic style in 1496. The Late Baroque façade was added in the 18th c. A commemorative tablet on the house calls to mind the distinguished composer, Bedřich Smetana who founded his first music school here. The wine bar, U Bindrů, which has traditions going back to the 16th c. is in the Early-Baroque house "To the Blue Star" (U modré Hvězdy; No. 25). The house next door, N. Kamenci is also a Romanesque building with Gothic extensions. Its Late Gothic doorway stems from the 16th c. and the Early Baroque façade from the 17th c.

A brass plate on the Old Town Square, inscribed in Latin and Czech, indicates the course of the meridian in relationship to Prague. This was used in the past for chronographic purposes.

Palace of Culture (Paláce kultury)   H5

The ultra-modern Palace of Culture (1981), a little way east of the Vyšehrad (see entry), is a fine example of contemporary Czech architecture. With its area of almost 280,000 sq. m (3,000,000 sq. ft) on several floors, it can accommodate a great variety of events

Location
Praha 4, Nusle

Metro Vyšehrad

Palace of Culture

– mass rallies, concerts, theatrical performances including Laterna Magica productions, exhibitions, entertainments, dances. A footbridge leads to the Hotel Forum on the opposite side.

Congress Hall

The central element in the building is the Congress Hall, which can be converted into an auditorium for an audience of up to 3000. The mighty organ is a reminder that the hall can also be used for concerts (e.g. during the Prague Spring Festival).

Other facilities

For smaller events there are four smaller halls, equipped with every technical refinement. There is also a hall specially designed for exhibitions as well as a number of conference rooms of different sizes.

Restaurant, etc.

Within the Palace of Culture are the select Panorama Restaurant and the elegant Vyšehrad Café, both affording fine views. An exclusive rendezvous for night-life enthusiasts is the Krystal night club.

Petřín Hill

E 1/2 F2

Location
Malá Strana, Praha 1

Trams
9, 12, 22

From the gardens of Strahov Abbey (see entry) a path climbs up through the old seminary gardens to the top of Petřín Hill; alternatively there is a cable-car. The house next to the stopping point was formerly occupied by a vine-growing family and was converted into the Nebozízek café in 1984/5. It takes its name from the vineyard mentioned in records as far back as 1433. There is a magnificent view of Prague from the top of the hill.

"Hunger" wall

Hanau Pavilion (p. 112)

Red Army Memorial

Petřín Hill

On the hill – an eastern outlier of the White Mountain (see entry) – are the Petřín Tower, St Lawrence's Church, the People's Observatory, the Mirror Maze and the Hunger Wall.

Petřín Tower
(Petřínská rozhledna)

This 60 m (200 ft) high tower, made of iron, was erected for the Prague Industrial Exhibition of 1891 on a model of the Eiffel Tower, and now serves as a television tower. The new television tower being constructed in the east of the town should be completed by 1992. From the upper gallery (384 m (1260 ft) above sea-level) there are far-ranging scenic views of Prague and Central Bohemia.

St Lawrence's Church
(Kostel svatého Vavřince)

Originally Romanesque, St Lawrence's Church (first recorded in 1135) was rebuilt between 1735 and 1770 (architect I. Palliardi) in Baroque style as a domed church with two towers. The statue of St Adalbert (1842) on the outside wall is by F. Dvořáček. Above the main altar is a painting by J. C. Monnos (1693) depicting the saint's torture. The legend of the foundation of the church on the site of a pagan place of worship in the year 991 is depicted on the ceiling fresco (1735) in the sacristy. The German name of Petřín Hill, Laurenziberg or St Lawrence's Hill is derived from the patron saint of the church, i.e. St Lawrence.

Pavilion

Near the church is a pavilion containing a panorama of the Prague students' fight against the Swedes in 1648 (by Karl and Adolf Liebscher and V. Bartoněk, 1898).

Mirror Maze
(Bludiště)

A small wooden building which formed part of the old Charles Gate in the Vyšehrad (see entry) now houses the Mirror Maze (Bludiště; open Apr.–Oct., daily 9 a.m.–7 p.m.), which was constructed at the same time as the Petřín Tower.

People's Observatory

The first section of the People's Observatory (Hvězdárna hlavního města Prahy), designed by V. Veselik and built on St Lawrence Hill, was opened to the public in the summer of 1928. It now houses the Astronomical Institute which is a department of the Czech Academy of Science. Like the Planetarium the observatory also stages a series of astronomical events and conducted tours, exhibitions as well as the very popular Wednesday evening social gatherings take place here. In addition, courses are held on astronomy and cosmology and there are also lectures on geography and special school courses. The modern instruments include a 40 cm Zeiss telescope but the oldest, large telescope known as "The King" and the original "Comet-finder" lightweight telescope are also in use today.

The People's Observatory is open in the evening, apart from Mondays, to allow amateur astronomers to pursue their hobby. Opening times: Jan., Feb., Oct.–Dec. every Tues.–Fri. 6–8 p.m.; Sat., Sun., 10 a.m.–12 noon; 2–5 p.m.; 6–8 p.m. March, Sept., Tues.–Fri. 7–9 p.m.; Sat., Sun., 10 a.m.–12 noon; 2–6 p.m.; 7–9 p.m. April, Aug., Tues.–Fri. 2–6 p.m.; 8–10 p.m.; Sat., Sun., 10 a.m.–12 noon; 2–6 p.m.; 8–10 p.m. May–July Tues.–Fri. 2–7 p.m.; 9–11 p.m. Sat., Sun., 10 a.m.–12 noon; 2–7 p.m.; 9–11 p.m.

Hunger Wall

From the top of the hill the old town wall, built in 1360–62 in the reign of Charles IV, runs down to the foot. It is called the Hunger Wall from a legend that it was built to provide employment in the fight against hunger.

Memorial to Soviet Tank Corps

At the foot of St Lawrence Hill is the square dedicated to the Soviet Tank Corps (Náměstí sovětských tankistů). A monument serves as a reminder of the liberation of Prague on 9 May 1945 by the Tank Corps under the command of General Leljuschenko. No. 23 tank which stands on the plinth is said to have been the first to enter Prague.

Pinkas Synagogue

See Josefov

Poděbrady

Poděbrady (Podiebrad, 187 m (592 ft) above sea-level; pop. 12,000). Situated on the Elbe (Labe), Poděbrady Spa is known for its alkaline and mineral thermal springs which have been effective in the treatment of vascular and heart disorders since 1907. The Bohemia glassworks have achieved international recognition for their products made from Bohemian lead crystal. A Renaissance castle (16th c., rebuilt in the 18th and 19th c.), used by Ferdinand I, Rudolf II and Marie Theresa as a hunting lodge, now stands on the site, in the main square, of an old Gothic fortification (13th c.). Opening times: May–Oct. Tues.–Sun. 9 a.m.–5 p.m. In front of the castle, there is an equestrian statue of the Bohemian King George of Poděbrady (Jiří z Poděbrady; 1420–71) who is supposed to have been born here. The Gothic church and the tiny Bergmann Church on the left bank of the Elbe are also of interest.

Location
48 km (30 miles)
east of Prague

Podiebrad Palace (Poděbrady palác)   E4

One of the best preserved Romanesque houses in Prague can be seen in Chain Street (Řetězová). The palace was built at the end of the 12th c. or at the beginning of the 13th c. for the Lords of Kunštát and Poděbrady and the cross vaulting in the basement is all that remains of the original Romanesque palatial building. Records show that the palace was owned in 1406 by Lord Boczko of Kunstatt who was the uncle of George of Poděbrady (1420–71) and in the mid 15th c. the Viceroy of the Kingdom of Bohemia. After his election in 1458 King Poděbrady, often described in history books as the Hussite king, moved to the royal palace which then stood in what is now known as the Square of the Republic. The palace was pulled down in 1902. Since that time Podiebrad Palace changed hands several times and in 1970 the restored building was opened to the public. A room dedicated to George of Podiebrad was opened here on the occasion of the 525th anniversary of his election as King of Bohemia. The Prague Centre for the Preservation of Ancient Monuments has a display in the basement of archaeological finds relating to the earlier history of Prague including ceramics from the 9th c., tiles from the 14th c., weapons used by the Hussite armies and tools from the Gothic period.

Location
Praha 1, Staré Město,
Řetězová 3

Metro
Můstek

Trams
17, 18

Opening times
May–Sept. Tues.–Sun.
10 a.m.–6 p.m.

Portheimka   G3

This Baroque mansion was built by Kilian Ignaz Dientzenhofer in 1729 for his own family. In 1758 it was taken over by Count Franz Buquoy and later, in the 19th c. passed into the hands of a Prague

Location
Třída S. M. Kirova 12,
Smíchov, Praha 5

industrialist named Portheim. The ceiling-frescoes of "Bacchus' Feast" (1729) in the central room are by V. V. Reiner. Part of the house was pulled down in 1884 to make room for St Wenceslas's Church.

The house is now occupied by Gallery D, where occasional exhibitions are mounted.

*Powder Tower (Prašná brána)   E5

This 65 m (215 ft) high Late Gothic tower, through which the trade route from the south entered Prague, was modelled on the Old Town Bridge Tower (See Charles Bridge) and formed part of the fortifications of the Old Town. The tower, successor to an earlier 13th c. fortified gateway, was built (1475 onwards) by M. Rejsek for King Vladislav Jagiello at the behest of the Municipal Council. After Vladislav who lived initially in the adjoining royal courtyard (no longer in existence) moved his residence to the Hradčany (see entry) the importance of the Powder Tower declined.

Location
Náměstí Republiky (Republic Square), Staré Město, Praha 1

Metro
Můstek
Náměstí Republiky

Trams
5, 10, 24, 26

The tower acquired its present name in the 18th c., when it was used as a powder-magazine. The sculptural decoration was badly damaged during Frederick the Great's siege of Prague in 1757. In 1875 the tower was reconstructed in neo-Gothic style by J. Mocker and given a new steeple roof and wall-walk. The sculpture – which includes portraits of Bohemian kings – was the work of J. Šimek, A. Wildt and other sculptors.
If the 186 steps can be attempted, there is a fine view of Prague from the top of the spiral staircase.

Opening times
Sat., Sun. and pub. hol.,
May–Sep. 10 a.m.–6 p.m.
Apr. and Oct. 10 a.m.–5 p.m.

Closed
Nov.–Mar.

Průhonice Gardens

The Renaissance palace of Průhonice, built in the 16th c. and remodelled at the end of the 19th c. contains an outstanding collection of plants. The park was laid out by the palace's former owner, Ernst Sylva-Taroucca and covers 260 ha. It is open to the public (Apr.–Oct. 7 a.m.–7 p.m.) although the palace itself is not. There is a horticultural research station in the park.

Location
16 km (10 miles)
south-east of Prague

*St Agnes's Convent (Klášter Anežsky)   D4

St Agnes's Convent, which now houses collections belonging to the National Gallery and the Museum of Applied Art, is one of the most important historical buildings in Prague and as such has been declared a national monument.
The convent was founded in 1234 by Agnes, a sister of King Wenceslas I, for the Order of Poor Clares (Franciscan nuns). Agnes later entered the Order and became the convent's first Abbess. The Minorite monastery which adjoins the convent was completed by 1240. Subsequently the churches of St Barbara (1250–80) and St Francis (c. 1250) and the conventual buildings were erected in the Burgundian Cistercian Gothic style.
The Church of St Salvator (1275–80) is one of the most significant examples of the Bohemian Early Gothic style. Recent surveys show that the church probably served as the burial place of the Přemysliden dynasty (sculpture on the capitals portraying the Přemysliden rulers). The presbytery and the Church of St Barbara,

Location
Anežská ulice, Staré Město,
Praha 1

Buses
125, 133, 144, 156, 187

Trams
3, 26, 29

Opening times
Tues.–Sun. 10 a.m.–6 p.m.

◄ *Powder Tower*

St Agnes's Convent
FIRST FLOOR

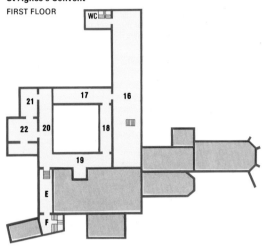

St Agnes's Convent
Bývalý klášter Anežský

National Gallery
19th c. Czech Painting
Museum of Applied Art
19th c. Bohemian Applied Art

GROUND FLOOR

© Baedeker

which was remodelled in Baroque style in 1689, date back to the 14th c. Archaeological excavations in the presbytery revealed the tomb of King Wenceslas I, and that of Agnes, the founder of the convent (d. 1282) as well as several Přemysliden tombs.

After the dissolution of religious houses in the reign of the Emperor Joseph II (1782) the convent fell into a state of disrepair. It has been restored in recent years after extensive preliminary archaeological investigation. As part of a further phase, the other churches and the old Minorite monastery were also restored by 1986. The National Gallery stages exhibitions here in the new rooms.

On the ground floor is an exhibition of Bohemian applied art, as well as 19th c. paintings depicting scenes from Czech history. There is a fascinating range of beautiful objects, from Rococo and Empire to neo-Renaissance and Jugendstil (Art Nouveau). Special mention must be made of the display of Bohemian glass.
Room 16 on the first floor, is devoted to 19th c. Czech painting. Of particular interest are the portraits by Antonín Machek (1775–1844), the landscapes of August Piepenhagen (1791–1868), the still-life paintings of Josef Navrátil (1798–1865) and an extensive display of the works of Josef Mánes (see Notable Peŕsonalities), the leading painter of the period among them, "Luise Bělská" (1857) and "The Seamstress" (1857/1859).
Also worth looking at are the works of the "National Theatre generation" (Rooms 17–20), including landscapes by Antonín Chittussi (1847–91) and Historical pictures by Václav Brožik (1851–1901). There are also paintings by Mikoláš Aleš (1852–1913) who produced the illustrated "Legend of the Fatherland" for the National Theatre.
Rooms 21 and 22 contain works by artists of the end of the 19th c. – genre pictures by Hanuš Schwaiger (1854–1912) for example "The Fish Market in Bruges" (1889), expressive paintings by Maximilian Pirner (1854–1924) including "The Lovers" (1885) and the still-life paintings, bathed in a poetic twilight, of Jakub Schikaneder (1855–1924) such as "Autumn" (1884).

St Barbara's Church

See St Agnes's Convent

St Cajetan's Church

See Thun-Hohenstein Palace

St Catherine's Church (Kostel svaté Kateřiny)   G5

This church, which formerly belonged to St Catherine's Convent is now used as an exhibition hall for the sculpture collection of the Municipal Lapidarium. When remodelling the church in 1737–41 F. M. Kaňka incorporated the original Gothic octagonal tower in the new Baroque structure. The slender form of the tower has led it to be called "Prague's minaret". The interior has frescoes by V. V. Reiner ("Life of St Catherine") and stucco-work by B. Spinetti. The conventual buildings are now occupied by the regional psychiatric clinic.

Location
Kateřinská ulice (entrance in Viničná ulice), Nové Město, Praha 1

Metro I. P. Pavlova

Opening times
In summer, Sat. 1 a.m.–6 p.m. Sun. 10 a.m.–6 p.m.

*St Clement's Church (Kostel svatého Klimenta)   E4

Location
Karlova ulice, Staré Město,
Praha 1

Metro
Staroměstská
Národní

Tram
17, 18, 21

The Baroque Church of St Clement, built between 1711 and 1715, is part of the Clementinum (see entry) complex and is linked with the Latin Chapel by an iron grille. The sculpture in the interior is among the finest Baroque sculpture in Bohemia. The eight figures of Evangelists and Fathers of the Church are by Matthias Bernhard Braun, as is the wood-carving on the side altars, the pulpit and the confessional. The altar-piece by Peter Brandl represents St Lienhard.

St Clement's is now used by the Greek Catholic community as a guest-house.

SS Cyril and Methodius, Church   F4
(Kostel svatého Cyrila a Metoděje)

Location
Resslova ulice, Nové Město,
Praha 1

Buses
6, 7, 176

Trams
3, 4, 16, 17, 18, 21, 22, 24, 27

This Baroque church, built by Kilian Ignaz Dientzenhofer in about 1740, was originally dedicated to St Charles Borromeo. In 1935 it changed its dedication on being taken over by the Czechoslovak Hussite Church. The interior has stucco decoration by M. I. Palliardi.

The crypt of the church was used as a hiding-place by the Czech paratroops who killed Reinhard Heydrich, the German "Protector" of Bohemia and Moravia, at Lidice in 1942. The Nazi authorities reacted with predictable brutality: Lidice was razed to the ground, all the male inhabitants over 16 were shot and the women were sent to concentration camps and the children to indoctrination camps. None of the Resistance fighters in the crypt survived. They are commemorated by a tablet bearing their names.

Diagonally across the street is the Church of St Wenceslas in Zderaz (see entry) and Charles Square (see entry) with its various features of interest (Faust House, St Ignatius's Church, New Town Hall) is only a few paces away.

St Francis's Church

See Knights of the Cross Square
See St Agnes's Convent

St Gallus's Church (Kostel svatého Havla)   E4

Location
Havelská ulice, Staré Město,
Praha 1
(Pedestrian zone)

Metro
Můstek

St Gallus's Church was founded in 1232, at the same time as the settlement of South Germans known as "St Gallus's Town", and was completed in 1263 to become one of the four parish churches of the Old Town. It was rebuilt in High Gothic style in 1353, and further remodelling in the 18th c. gave it its curving façade and twin towers.

The Baroque interior has fine altar-pieces and (on the left) a carved Pietà, probably by F. M. Brokoff. In the side chapel to the right is the tomb of the painter Karel Škréta.

From 1363 the Austrian preacher Konrad von Waldhausen, a forerunner of the Reformer Jan Hus, officiated in this church at the wish of Charles IV.

St George's Basilica

See Hradčany

St Giles's Church (Kostel svatého Jiljí)   E4

Originally Romanesque, St Giles's Church was rebuilt in Gothic style between 1339 and 1371. It belonged to the Hussite Utraquists (who believed that laymen should receive Communion in both kinds), and after the Battle of the White Mountain was presented by Ferdinand II to the Dominicans (1625).
The church was remodelled in Baroque style in 1733. V. V. Reiner was responsible for the ceiling-painting ("Glorification of the Dominican Order") and the altar-piece (St Wenceslas) in the chapel in the north aisle, and was himself buried in the church. The sumptuous confessionals were the work of R. Prachner.
Recitals of Church music are given here from time to time.

Location
Husova třída, Staré Město,
Praha 1

Metro
Staroměstská
Národní

Trams
9, 18, 22

St Henry's Church (Kostel svatého Jindřicha)   E5

This Gothic church was built in 1348–51 as the Parish Church of the New Town, which was founded at the same time. The tower was added in 1475 and originally served as part of the town's defences; it was restored in Gothic style in 1879. In front of the church is a statue of St John of Nepomuk.
The interior was remodelled in Baroque style in the 18th c. On the right of the high altar is a very fine panel-painting of the Virgin. The paintings in the Chapel of the Virgin Dolorosa ("Transfiguration", "Immaculata") are by V. V. Reiner, who ranks with P. J. Brandl as the finest of the fresco-painters of the Bohemian High and Late Baroque.

Location
Jindřišská ulice (corner of
Jeruzalémská), Nové Město,
Praha 1

Metro
Můstek

Trams
3, 9, 24

St Ignatius's Church

See Charles Square

*St James's Church (Kostel svatého Jakuba)   E4

St James's was built in 1232 as the church of the old Minorite friary (on the north side). After being destroyed by fire in 1366 it was rebuilt in Gothic style, and was given its present Baroque form between 1689 and 1739. The stucco front, with figures of SS James, Francis and Anthony of Padua, was the work of Ottavio Mosto.

Location
Malá Štupartská (corner of
Jakubská),
Staré Město, Praha 1

Metro
Můstek
Náměstí Republiky

Interior
St James's is most notable for its interior, with its delicately modelled pilasters and its 21 altars. It is Prague's longest church after St Vitus's Cathedral (see Hradčany), and its rich decoration makes it one of the most beautiful. The "Martyrdom of St James" on the high altar is by V. V. Reiner, the ceiling-frescoes ("Life of the Virgin", "Glorification of the Trinity") by F. Q. Noget. The Baroque monument of Count Vratislav Mitrovic was designed by Johann Bernhard Fischer von Erlach and executed by F. M. Brokoff (1714–16).

Trams
3, 5, 24, 26

St James's Church

"The Child Jesus" of Prague

Since the church has excellent acoustics it is frequently used for concerts and recitals.

Remodelled in the 14th c. in Gothic style and later in Baroque the old Minorite friary on the north side of the church is now an art school.

St John of Nepomuk (Kostel svatého Jana Nepomuckého)   D2

Location
Kanovnická ulice, Nové Město, Praha 1

Metro
Hradčanská

Trams 22, 23

From Hradčany Square Kanovnická ulice runs north-west to the Church of St John of Nepomuk, the first church built in Prague by Kilian Ignaz Dientzenhofer (1720–28). Much of the tower was destroyed in 1815.

The church has fine ceiling-frescoes by V. V. Reiner ("Glorification of Life", "Miracles of St John of Nepomuk") and altarpieces by M. Willmann and J. K. Liška (1701).

St John of Nepomuk on the Rock   H4
(Kostel svatého Jana Nepomuckého na skalce)

Location
Vyšehradská 18, Nové Město, Praha 1

Trams
4, 6, 7, 16, 18, 24

Built by Kilian Ignaz Dientzenhofer in about 1730, this church, on a centralised plan with twin towers and a double external staircase, is one of the finest Late Baroque churches in Prague. The fresco of the Ascension of St John of Nepomuk is by K. Kovář (1748), the wooden statue of the Saint on the high altar by Johann Brokoff; a bronze statue based on this model is on the Charles Bridge (see entry).

St Longinus's Chapel (Rotunda svatého Longina) F5

This round Romanesque chapel was originally a village church, but the village of Rybníček to which it belonged was absorbed into the New Town at some time after 1257 and is now recalled only by the name of the street in which the church stands. Until the 14th c. it was dedicated to St Stephen.

Notable features of the interior are the Baroque altar and the representation of the Crucifixion in which Longinus appears. According to an apocryphal source Longinus was the soldier, or captain, who pierced Christ's side with a lance. Legend has it that he later became a bishop in Cappadocia and suffered a martyr's death.

Location
Na Rybníčku,
Nové Město, Praha 1

Metro
Karlovo náměstí

Trams
4, 6, 16, 18, 22, 24

St Martin's Chapel

See Vyšehrad

St Mary of the Snows (Kostel Panny Marie Sněžné) E4

The Church of St Mary of the Snows, the construction of which was ordered by Charles IV in 1347 as a monastic church for the royal coronation, was originally designed to surpass St Vitus's Cathedral in size, but by 1397 only the 30 m (100 ft) high choir had been completed, with the fine Gothic doorway on the north side. From the 15th c. the building fell into disrepair. In 1611 the vaulting collapsed and the Franciscans replaced it by a Renais-

Location
Jungmannovo náměstí,
Staré Město, Praha 1

Metro
Můstek

St Mary of the Snows

sance ceiling. The Baroque high altar (1625–51) is the largest in Prague. Over the left-hand side altar is an "Annunciation" by W. L. Reiner. There is a fine pewter font of 1459.

The church played an important part in the history of the Hussite movement. Here Jan Želivský preached to congregations of the city's poor against the Papal Church, the nobility and the wealthy burghers; and it was Želivský who in 1419 stormed the New Town Hall with the most radical of his supporters and triggered off the Hussite Wars by throwing the Emperor's Catholic councillors out of the window. Even after the murder of Želivský in 1422 (he was buried in the church), St Mary of the Snows remained a centre of the Hussite movement.

Franciscan Garden
(Františkánská zahrada)

The Franciscan Garden on the south side of the church has been a public park since 1950. It is linked by shopping arcades with Wenceslas Square, Palackého and Vodičkova ulice.

Jungmann Square
(Jungmannovo náměstí)

In Jungmann Square is the Jungmann Monument (Jungmannův pomnik; by L. Šimek). The writer and philologist Josef Jungmann (1773–1847) played a leading part in the rebirth (obrozeni) of Czech national consciousness during the Romantic period. He also compiled a large German-Czech dictionary and wrote a history of Czech literature.

St Mary's of Emmaus

See Emmaus Abbey

St Mary the Victorious (Kostel Panny Marie Vítězné)   E3

Location
Karmelitská ulice, Malá
Strana, Praha 1

Trams
12, 22

In Karmelitská ulice (Carmelite Street), a short distance west of Maltese Square, stands the Early Baroque Church of St Mary the Victorious, originally a Carmelite church, which was built on the site of an earlier Hussite church after Ferdinand II's victory in the Battle of the White Mountain.

The interior is modelled on the Gesú Church in Rome. On the right-hand wall is the "Christ Child of Prague", a wax figure just under 50 cm (20 in) high, originally from Spain, which Princess Polyxena Lobkowitz presented to the Carmelite friary in 1628 and which is still much revered. The altar, by F. M. Lauermann (1776) has statues by P. Prachner and a silver casket depicting the Christ Child (1741) by J. Pakeni. The high altar is from the studio of J. F. Schor. In the catacombs under the church are the dried-up bodies – well preserved because of the circulation of air – of Carmelite friars and their benefactors (closed to the public on health grounds).

St Mary under the Chain (Kostel Panny Marie pod řetězem)   E3

Location
Lázeňská ulice,
Malá Strana, Praha 1

Trams
12, 22

St Mary under the Chain is the oldest church in the Lesser Quarter (see entry), founded in 1169 together with the house of the Knights of Malta, the administrative centre of the Order in Bohemia. In the right-hand wall of the forecourt can be seen the remains of the original Romanesque church which was burnt down in 1420. The church's two massive towers were completed in 1389. The presbytery was given its present Baroque form by Carlo Lurago in the 17th century.

Notable features of the Baroque interior are the altar-piece on the high altar ("The Assumption of the Blessed Virgin") and a painting of St Barbara, both by Karel Škréta.

St Nicholas's Church in the Lesser Quarter

See Lesser Quarter Square

St Nicholas's Church in the Old Town (Kostel svatého Mikuláše)   E4

St Nicholas's, originally the church of a Benedictine house (see Emmaus Abbey), now belongs to the Czechoslovak Hussite Church.

Location
Staroměstské náměstí
(Old Town Square),
Staré Město, Praha 1

Situated on the north-west corner of the Old Town Square, this Baroque church with its monumental south front, long nave with side chapels and dome was built by Kilian Ignaz Dientzenhofer in 1732–35. The sculptural decoration is by Anton Braun, the rich stucco-work by B. Spinetti, the ceiling-paintings ("Lives of St Nicholas and St Benedict") by Peter Asam the Elder who was also responsible for the frescoes in the presbytery and side-chapel. The crystal chandelier in the nave is from the Harrov glassworks (end of the 19th c.). The statue of St Nicholas on the lateral façade is by B. Šimonovský (1906).

Metro
Staroměstská

Buses
133, 144, 156, 187

Tram
17

After the dissolution of the abbey, the high altar, pews and many of the pictures were removed to other churches.

Adjoining the church, to the left, is the house in which the Prague writer Franz Kafka was born (bust). At that time, his father's business was opposite St Nicholas's Church in the Kimsky Palace.

Franz Kafka's birthplace

SS Peter and Paul

See Vyšehrad

St Salvator's Church (Kostel svatého Salvátora)   E4

St Salvator's, originally a German Lutheran church, now belongs to the Bohemian Brethren.

Location
Salvátorská,
Staré Město, Praha 1

Designed by the Swiss-born architect, J. Christoph, the church was built in Renaissance style in 1611–14. After being acquired by the Paulian Order – a closed order of St Francis of Paola, it was remodelled in Baroque style in the mid 17th c. and provided with a tower in 1720. The work was financed with the help of contributions from all over Protestant Europe.

Metro
Staroměstská

St Salvator's Church in the Clementinum

See Knights of the Cross Square

St Stephen's Church (Kostel svatého Štěpána)   F5

Location
Štěpánská ulice,
Nové Město, Praha 1

Metro
I. P. Pavlova

Trams
4, 6, 16, 22

St Stephen's was founded by Charles IV in 1351 as the Parish Church of the upper New Town and completed in 1394. The tower was added at the beginning of the 15th c. The Gothic exterior of the church has been preserved in spite of restoration work in 1876 and 1936.

Notable features of the church, in addition to the Baroque interior, are the Gothic font made in pewter by B. Kovář (1462), the Gothic Madonna (1472), the Late Gothic stone pulpit (15th c.), three paintings by Karel Škréta ("St Rosalia", c. 1660, second pier on right; "St Wenceslas", c. 1650, north side of choir; "Baptism of Christ", 1649, end of north aisle) and the monument of the Baroque sculptor M. B. Braun (1684–1738).

St Thomas's Church (Kostel svatého Tomáše)   D3

Location
Letenská ulice, Malá Strana,
Praha 1

Metro
Malostranská

Trams
12, 22

St Thomas's the Gothic origin of which is made evident by the massive buttresses on the outer walls of the presbytery, was founded in 1285 for the Order of Augustinian Hermits and was completed, with the Augustinian friary (now an old people's home) and St Thomas's Brewhouse, in 1379.

The church was remodelled in Baroque style by Karl Ignaz Dientzenhofer (1727–1731). In a niche over the Renaissance doorway (by Campione de' Bossi, 1617) can be seen statues of St Augustine and St Thomas by Hieronymus Kohl (1684).

In the richly appointed interior are paintings and statues by Bohemian artists. Wenzel Lorenz Reiner painted the ceiling fresco (1730) in the nave with scenes from the life of St Augustine. The paintings in the choir and in the dome of the Legend of St Thomas are by the same artist. The high altar (1731) with figures of the saints by J. A. Quitainer, F. M. Brokoff and I. Müller, is the work of Karel Škréta who was also responsible for the paintings of St Thomas (1671) on the altar in the transept and "The Assumption of Our Lady" (1644) in the choir.

St Ursula's Church (Kostel svaté Voršily)   F4

Location
Národní třída 8, Nové Město,
Praha 1

Metro
Národní

Trams
9, 18, 21, 22

The Baroque Church of St Ursula, with a striking richly articulated façade was built in 1702–04 to the design of Marcantonio Canevale; it formed part of the Ursuline convent. The statuary group of St John of Nepomuk is by Ignaz Platzer the Elder (1747).

Particularly notable features of the sumptuous Baroque interior are the ceiling-frescoes by J. J. Steinfels ("Holy Trinity", 1707) and the altar-piece ("Assumption") by Peter Johann Brandl.

There is an excellent wine-bar, the Klášterni Vinárna, in the former conventual buildings.

St Vitus's Cathedral

See Hradčany

St Thomas: ceiling painting by W. L. Reiner ▶

St Wenceslas's Church in Zderaz   F4
(Kostel svatého Václava na Zderaze)

Location
Resslova ulice, Nové Město, Praha 1

Trams
4, 6, 16, 22

Metro
Karlovo náměstí

St Wenceslas's was originally the Parish Church of the commune of Zderaz, which was later incorporated in the New Town. It has belonged to the Czechoslovak Hussite Church since 1926.

The present 14th c. Gothic church has preserved some remains of a Romanesque nave and tower. The choir has fragments of Gothic wall-paintings of about 1400. The Late Gothic stellar vaulting by K. Mělnický dates from 1587.

Schönborn Palace (Schönbornsky palác)   E2

Location
Tržiště 15, Malá Strana, Praha 1

Trams
12, 22

This extensive palace, with four wings enclosing a courtyard and a relatively plain façade, is now the United States Embassy. The pediments and dormer windows and the four statues of giants at the entrance to the courtyard were added by Giovanni Santini in about 1715.

The gardens of Schönborn Palace were already widely famed in the mid 17th c. They rise in terraces from the formally patterned flower-beds to the arcaded pavilion (formerly a wine-press) at the top of the hill, from which there is a superb view of Prague.

Schwarzenberg Palace

See Hradčany Square

Šitek Water-Tower (Šitovská věž)   F3/4

Location
Gottwaldovo nábřeží 250, Nové Město, Praha 1

Trams
3, 7, 17, 21

Near the Mánes Exhibition Hall (see entry) is a Renaissance tower, originally dating from the mid 15th c. but much damaged by fire and bombardment and frequently restored. The Baroque roof was added at the end of the 18th c. The tower is named after the owner of the mill (b. 1451). It has supplied the New Town with water since the end of the 15th c.

*Slapy Dam (Slapská přehradní nádrž)

Location
24 km (15 miles) south of Prague

Surrounded by wooded hills, Vltava Reservoir, which is very long and with many bays, is one of the favourite health and leisure areas of Prague. The dam (Slapská přehradní nádrž), situated approximately 5 km (3 miles) east of the village of Slapy was built between 1949 and 1954. It is a little less than 70 m (230 ft) high, 260 m (284 yd) long along the top and, with a surface area of roughly 1400 ha. (3459 acres) provides some 270 million cubic metres (353 cubic yards) of water for the production of energy. The 44 m (48 yd) long reservoir which can be reached by boat from Prague during the summer offers ample opportunity for all types of water sport, fishing and round trips on one of the Vltava boats which moor either close to the dam itself or in Nová

Rabyně. Like Živohoštvo, this resort is very popular with tourists. For those interested in archaeology, a trip to Hrazany would be recommended as there is a Celtic settlement there which dates back to 1 B.C.

Smetana Embankment (Smetanovo nábřeží)   E/F3

The Smetana Embankment, named after the composer Bedřich Smetana (see Notable Personalities), extends from the National Theatre to a small peninsula just before the Charles Bridge, offering magnificent views of the Hradčany and Charles Bridge (see these entries).

Location
Staré Město, Praha 1

Metro
Staroměstská

Tram
17, 18

A path runs along the little peninsula past the Emperor Francis I Monument to the 15th c. water-tower of the Old Town and the Smetana Museum, which has been housed since 1936 in the old Prague Waterworks building, a neo-Renaissance structure of 1883. The sgraffiti decoration by J. Šubič is interesting. On display in the museum are original scores of music, musical instruments, letters and other personal items which belonged to the famous composer. Concerts and recitals are held in the large hall. The Smetana Memorial, the work of J. Malejovský, was unveiled in 1984 and stands in front of the museum on Novotný Steg.

Spanish Synagogue

See Josefov

Smetana Museum in the former Prague waterworks

Star Castle

See White Mountain

Sternberg Palace

See European Art Collection of the National Gallery

**Strahov Abbey and National Literary Memorial E1
(Strahovský klášter, Památník národního písemnictví)

Location
Strahovské nádvoří 132
(entrance also at Pohořelec 8),
Hradčany, Praha 1

Trams
22, 23

Opening times
Tues.–Sun. 9 a.m.–5 p.m.

Strahov Abbey, Prague's second oldest religious house, was built in 1143 by Duke Vladislav II, at the request of the Bishop of Olomouc (Olmütz), Jindřich Zdik, for the Premonstratensian Order. The name "Strahov" is derived from the abbey's position on a hill overlooking the Lesser Quarter and the entrance to Prague Castle (strahovati – to keep watch over). Following a fire in 1258 which destroyed the first library, the abbey was rebuilt in Gothic style. Originally situated outside the city gates, it was brought within the walls by Charles IV in 1360. The Hussite Wars put an end to any further building in Gothic form during the 15th c. but once these were over, the abbey enjoyed a golden age under Abbots Jan Lohel, Kašpar of Questenberk and Kryšpin Fuck who rebuilt the complex on a grand scale in Renaissance style. At the end of the Thirty Years' War, the building was completely devastated by Swedish soldiers. After peace came to Westphalia it was possible, as so many more books had been

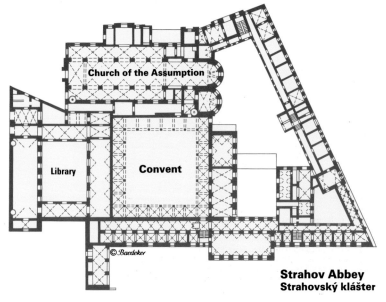

Strahov Abbey
Strahovský klášter

View of Strahov Abbey from Hradčany

Strahov Abbey: Philosophical Library

acquired, for a new library to be built in 1671, the so-called Theological Library. Baroque alterations were made to the abbey grounds between 1682 and 1689 under the direction of the distinguished architect J. B. Mathey. Extensive parklands, gardens and orchards enhanced the overall view. During the Wars of the Austrian Succession the abbey was again badly damaged in 1741 and the repair work lasted 40 years. Work on the building, in Classical style, ended with the Philosophical Library, the most famous of Prague's monuments of that era. The remains of the original Romanesque foundations were uncovered when restoration work was being carried out in the grounds during 1950/54.

The abbey courtyard can be entered either through a passage at Pohořelec 8 (stairs) or, preferably, by a short steep street at the west end of the square, turning left through a Baroque gateway (1742), over which there is a statue (1719) by J. A. Quittainer of St Norbert, the founder of the Premonstratensian Order.

Abbey grounds

In the courtyard, on the left, is the old Chapel of St Roch (Kaple svatého Rocha, 1603–11), now the Musaion exhibition hall. Rudolf II had the chapel built in 1603–1617 as a gesture of thanksgiving for being spared from the plague of 1599. Straight ahead behind St Norbert's Hall (17th c.), is the 17th c. Church of the Assumption (Kostel Nanebevzetí Panny Marie), with three naves and a sumptuous Baroque interior dating back to the mid 18th c. (restored during the 1970s). Stucco work by M. I. Palliardi; pictures of the Virgin Mary by J. Kramolin and I. Raab; the high altar (1768) by J. Lauermann with a relief portraying saints by I. F. Platzer.

In the Pappenheim Chapel in the south aisle is the Tomb of Gottfried Heinrich zu Pappenheim (1594–1632), a cavalry general who fell in the Battle of Lützen.

Adjoining the church are the conventual buildings, some of which date from the Romanesque period. Together with the library and the cloister they have housed the National Literary Memorial and the Museum of Czech Literature since 1953.

National Literary Memorial

The central element in the Museum of Czech Literature is the old abbey library (reached from the cloister by a staircase leading up to the first floor). At the beginning of the '60s the literary archives of the National Museum (the legacy of authors, scientists and intellectuals) were transferred to this collection. The Alois-Jirásek and Mikláš-Aleš Museums housed in neighbouring Star Castle also became affiliated to the library. Among the priceless items which include 130,000 volumes (400,000 housed at Kladruby) representing literary development from the 9th–18th c., there are 2500 early printed books, 5000 manuscripts and numerous old maps. The extensive works, readily placed at the disposal of the learned public, persuaded Emperor Joseph II among others at the end of the 18th c. to protect the abbey community.

Libraries
The two finest rooms in the museum are the Theological Library with rich stucco ornament and paintings of 1723–27 by the Strahov monk Siard Nosecký; (its barrel vaulting by Giovanni Domenico Orsi of Orsini was built in Early Baroque style), and the

Theological Library

Philosophical Library in the neo-Classical west wing (by Ignaz Palliardi, 1782–84).

When decorating the Theological Library Nosecký was inspired by biblical quotations and by "De typho generis humani", the work of Abbot Hieronymus Hirnheim (1670–79). Its 25 frescoes symbolise the struggle for wisdom in relation to the love of knowledge and literature. A self-portrait of the artist may be seen in a window niche on the right. Placed alternately along the longer axis of the library are geographical and astral globes, three of which were made in the workshops of the Dutch cartographer, Wilhelm Blaeus. The Philosophical Library's dimensions (32 m (105 ft) long, 10 m (33 ft) wide, 14 m (46 ft) high) were designed to accommodate the richly carved bookcases (J. Lachhofer) from Bruch Abbey in Southern Moravia. The huge ceiling fresco by Franz Anton Maulperťsch (1724–96) of Langenargen (Lake Constance) depicts scenes from the intellectual history of mankind in the allegorical style of the Vienna Academy. There is an old book-case standing in the middle of the library which holds the botanical work, "Les Liliacées" in six volumes and "Le Musée Français" consisting of 4 volumes of essays. These were both presented to the library in 1812 by Empress Marie Louise of France. The bust of Emperor Franz I was made by F. X. Lederer (c. 1800).

Contents of the Library
Among the most valuable manuscripts in the library is the Strahov Gospel Book (9th/10th c.), a work of art in Ottonian Renaissance style by artists of the Trier School. This Latin document was written in gold script on 218 sheets of parchment and later decorated with four ornate pictures of the Evangelists. Together with the St Markustorsi and the famous Kodex of Vyšehrad, it is one of the oldest manuscripts still in existence in Central Europe. Special mention must also be made of the Historia Anglorum chronicles, the accounts of the Italian campaign by Friedrich Barbarossa, the partially preserved Dalimil chronicles, the so-called Doxan Bible, the Late Gothic Schelmenberg Bible (Bishop Albrecht of Sternberg from the time of Charles IV) as well as scripts by Tomaš of Štitné and Jan Hus. Included among works dating back to the 15th c. are the Strahov Herbarium, a Latin dictionary and the medical books which once belonged to Dominie Ambrož. The 16th and 17th c. are represented by works of the Utraquists and Brotherly Unity and by catholic literature. Accounts of journeys and oceans, works by alchemists, essays on astronomy by Tycho Brahe, Kepler and Copernicus, oriental manuscripts and other literary curiosities are all included in the never ending list of items.

The museum also contains large stocks of books from many Bohemian religious houses dissolved after the Second World War.

Czech Scripts
In the cloister and adjoining rooms are approximately 50,000 examples of Czech literature of pre-Hussite and Hussite times, with particular emphasis on the latter, and also of the period of national revival in the 19th c. Those interested in printing, should also take a look at the old 17th c. printing press (reconstructed in 1953).

In addition to literature, a department of graphic arts and other works has been opened, and poetry evenings, concerts and special exhibitions are regularly held in the Lecture Room and the Exhibition Hall.

Stromovka Park

Location
U sjezdového paláce,
Holešovice, Praha 7

Trams 5, 12, 17

Opening Times
Tues.–Sun. incl.

This beautiful park extends from Letná Hill (see Letná Gardens) to the Vltava. On the south-west side of the park is an old hunting-lodge, originally dating from the 15th c. but rebuilt in neo-Gothic style in 1804, which now houses the newspapers and periodicals department of the National Museum.

Julius Fučík Park of Culture and Recreation

At the end of the 19th c. an exhibition site, designed by A. Wiehls, was formed at the eastern end of the park for the Jubilee Exhibition of 1891 and the 1895 Ethnographic Exhibition. The Prague model fair has been held here since 1918. At the beginning of the 50s the grounds were turned into the present day Julius Fučík Park of Culture and Recreation (Park kultury a oddechu Julia Fučíka), named after the left-wing journalist and author Julius Fučík who was arrested in 1943 by the Gestapo and tortured to death.

Palace of Congress

The entrance to the park is flanked by twelve fountains with animal motifs by J. Kavans. The iron structure of the Palace of Congress (Sjezový palác) erected at the turn of the century by B. Münzenberger and J. Fantas is very impressive. Alterations to the building were carried out between 1952 and 1955 by P. Smetana. Until the Palace of Culture was opened, most of the important conventions and meetings were held here.

Julius Fučík Park

The Prague Pavilion, the front of which is decorated with pictures by G. Zoula of famous people from Bohemian history as well as allegorical sculptures by F. Hergestell was also built at the turn of the century. It now houses the Lapidarium of the National Museum and features a remarkable collection of architectural artefacts and sculpture from the 11th to the 19th c.

Lapidarium

The circular "Pavilion with the panorama of the Battle of Lipany" was built in 1908 to the design of J. Koulas. There is a painting inside by L. Marolds (1898) portraying the Hussite battle which took place on 30 May, 1434.

Lipany Pavilion

Lectures and exhibitions covering various aspects of astronomy as well as literary and musical events are held on a regular basis in the domed building of the Planetarium, built between 1960 and 1962.

Planetarium

At approximately the same time as the Planetarium was put up, the Sports Hall, which can accommodate 18,500 spectators, was erected on the site of the old Mechanical Engineering Pavilion (1907). Concerts are held here in addition to sporting events.

Sports Hall

The large swimming pool was completed in 1976.

The Exhibition Pavilion, designed by F. Cubr, J. Hrubý and Z. H. Pokorný was awarded the first gold star and two prizes for architecture at the World Fair Expo 58 held in Brussels in 1958.

Exhibition Pavilion

Other features of the Park of Culture are an open-air theatre and a 3-D cinema.

Sylva-Taroucca Palace

See Na Příkopě

Terezín (Theresienstadt)

Situated on the banks of the Eger (Ohře), which flows into the Elbe a little further downstream, is the fortified town of Terezín (Theresienstadt; 155 m (345 ft) above sea-level; pop. 3000), built by Marie Theresa, i.e. Joseph II. Within just 10 years, a Bohemian model example emerged of a late 18th c. town planned in the Empire and Classic styles. Since the middle of the 19th c., the fortress, designed by General Pellegrini, served as an Austrian state prison in which, among others, the Sarajevo murderer, Gavrilo Princip was incarcerated. During the Second World War, the inhabitants of the town were driven out by the Nazis and Terezín was turned into a ghetto. From 1940 onwards, over 140,000 Jews from all over Europe were brought here before being transported to the gruesome extermination camps, predominantly Auschwitz. The so-called Small Fortress (Malá Pevnost) was also converted into a concentration camp. After the war a memorial ("Památnik Terezín") and a museum were erected here. A huge national cemetery stretches in front of the entrance to the fortress. The Jewish Cemetery is close to the old graveyard south of the town and from here the Menora Memorial (the seven-armed candlestick of the Jewish liturgy) dedicated to the Jewish victims of the Nazis, can be seen from afar.

Location
60 km (36 miles) north-west of Prague

Theatine Church

See Thun-Hohenstein Palace

Three Ostriches House (U tří pštrosů)   E3

Location
Dražického náměstí 12, Malá
Strana, Praha 1

Metro
Malostranská

Trams
12, 22

This handsome Renaissance house at the Charles Bridge (see entry), built in 1597, preserves remains of painting (by Daniel Alexius Květná, 1606) on its façade. The upper storey, in Early Baroque style, was designed by Caril Geer (1657).

The beamed ceilings in the rooms of the old inn (now a hotel) date from the 17th c.

Prague's first coffee-house was opened here in 1714 by an Armenian named Deodatus Damajan.

Thun-Hohenstein Palace (Thun-Hohenštejnský palác)   E2

Location
Nerudova ulice 20, Malá
Strana, Praha 1

Metro
Malostranská

Trams
12, 22

This Baroque palace was built by Giovanni Santini in 1710–25 for Norbert Vinzenz Kolowrat. It is now the Italian Embassy.

The palace, which is joined to the Slavata Palace in Thunovská Street, has a magnificent doorway with two heraldic eagles with outstretched wings (the device of the Kolowrat family) and figures of the Roman gods, Jupiter and Juno.

A short distance away, in the direction of the Hradčany (see entry), is St Cajetan's Church, also known as the Theatine Church,

Thun-Hohenstein Palace

Rococo gateway of Nostiz Palace

which was built between 1691 and 1717 to the design of Jean-Baptiste Mathey and Giovanni Santini.

Town Hall of the Old Town (Staroměstská radnice) E4

The former Town Hall of the Old Town is now used for cultural and social occasions (e.g. weddings), but has preserved its name as the Town Hall.

The history of the Town Hall, the oldest parts of which date from the 11th c., is a story of continuing building activity, involving both the conversion of existing burghers' houses and new construction.

In 1338 King John of Luxemburg granted the citizens of the Old Town the right to build their own Town Hall. The nucleus was a house belonging to the Stein family, to which a square tower was added in 1364. The oriel chapel on the north-east side of the tower was consecrated in 1381; it was badly damaged in 1945 but was restored after the war. A casket built into the wall contains earth from the Dukla Pass, where Czech and Russian forces drove back the Germans in 1944. Set into the paving in front of the east side is a stone commemorating the leaders of the Czech Protestant rising, executed in 1621: two white swords, crossed, with the Crown of Thorns, the date of execution and 27 small crosses. The bust, by K. Lidický is of the Hussite minister, Jan Želivský who was executed in 1422.

The Astronomical Clock on the south side of the tower was installed at the beginning of the 15th c. The Gothic doorway on the south front, the main entrance, was completed in 1480.

About 1360 the Kříž House was purchased to provide additional accommodation. The inscription over the Renaissance window, "Praga caput regni" (Prague, capital of the kingdom) dates from 1520.

In 1458 a third house was acquired, the Mikeš House, which was rebuilt in neo-Renaissance style in 1878.

In 1830 the House of the Cock (U kohouta) was added to the Town Hall complex. A Romanesque room is still preserved in the basement. Fine Renaissance ceilings and wall-paintings on the first floor.

Building activity did not come to an end until the end of the 19th c. The Town Hall was badly damaged on the last day but one of the Second World War, when the tower was bombarded by the remnants of the Nazi army and the neo-Gothic east and north wings and the municipal archives were destroyed. A small park occupies the site today.

The south wing was completely restored between 1978 and 1981 and the Council Chamber on the second floor has been preserved in its original Gothic form (1470).

In the Great Hall can be seen two pictures by the Czech Historical painter Václav Brožik, "The Election of George of Poděbrad as King of Bohemia" and "Jan Hus before the Council of Constance". The Wedding Room contains paintings by C. Bouda. In the cloister is the Municipal Gallery.

Location
Staroměstské náměstí (Old Town Square), Staré Město, Praha 1
Metro
Staroměstská

Opening times
Mar.–mid Oct., 8 a.m.–6 p.m.;
Mid Oct.–Feb., 8 a.m.–5 p.m.

Town Hall of the Old Town

There is a superb panoramic view of Prague from the 70 m (230 ft) high tower (restored 1984–89) the top of which may be reached by lift from the third floor.

House of the Minute (U minuty)

Adjoining the Town Hall on the south stands the House of the Minute, with sgraffito decoration (Biblical and mythological scenes); originally built about 1600, the house was later re-modelled in Renaissance style. The statue of a lion at the corner is 18th c. In the arcading is a passage leading into Little Square (Malé náměstí).

Astronomical Clock (Orloj)

"On this clock were to be seen the course of the heavens throughout the year, with the tale of the months, days and hours, the rising and setting of the stars, the longest and the shortest day, the equinoxes, the feast-days for the whole year, the length of day and night, the new and the full moon with the four quarters, and the three different hours of striking according to the whole and the half hour." So wrote the painter and engraver Matthäus Merian in 1650 about the astronomical clock on the south side of the Town Hall tower; and for the last 500 years hardly anything has changed in the appearance of the clock.

The clock was originally installed in 1410, but in 1490 it was rebuilt by one Master Hanuš of the Charles University. Legend has it that the Municipal Council then had him blinded to prevent him from constructing a similar marvel for any other town. Then, it is said, the blind man climbed the tower shortly before his death and stopped the clock. Thereafter the clock remained silent until

"Praga Caput Regni"

Astronomical Clock: Procession of apostles

Jan Táborský restored the mechanism to working order between 1552 and 1572.

The clock consists of three parts – the procession of Apostles, the face which tells the time and the calendar. The main attraction is the procession of the Apostles, which takes place every hour on the hour. Death, represented by a skeleton, pulls the rope of the funeral bell with one hand and raises his sand-glass in the other. The windows open, and Christ and the Twelve Apostles appear. After the windows have closed again a cock flaps his wings and crows, and the clock strikes the hour. Other characters who also feature in the scene are a Turk shaking his head, a miser gloating over his sack of gold and a vain man contemplating his face in a mirror.

The calendar was painted by Josef Mánes. The original is in the staircase hall of the Municipal Museum (see entry).

Troja Palace (Letohrádek Troja)   A3

This handsome Baroque palace, built by Jean-Baptiste Mathey in 1679–85, lies to the north of the Stromovka Gardens (see entry) in the district of Troja, on the far side of the Vltava.

The fine staircase in front of the house was a later addition; its sculptural decoration, depicting a fight between giants and Titans, was the work of Johann Georg and Paul Heermann of Dresden and the brothers Johann Josef and Ferdinand Maximilian Brokoff.

A notable feature of the interior is the Imperial Hall, with wall- and ceiling-paintings (1691–97) by the Dutch artist Abraham Godin. Mythological frescoes by the Italian artists, Giovanni and Giovanni Francesco Marchetti may be seen in the side-rooms of the palace and Early-Baroque vases and busts dating back to the 17th c. are on the terraces. Extensive restoration work was carried out on the palace at the end of the 1980s.

Location
Troja, Praha 7

Bus
112

Opening times
Apr.–Sep., Tues.–Sun.
9 a.m.–5 p.m.

Tuscan Palace (Toskánský palác)   B2

See Hradčany Square

Tyl Theatre (Tylovo divadlo)   E4

This neo-Classical building (architect A. Haffenecker), the first theatre in the Vltava Valley, was built for Count Anton von Nostitz-Rieneck in 1781–83. The east front and the interior decoration were the work of Achill Wolf (1881). At the end of the 18th c. the Nostitz Theatre was the theatre patronised by the nobility; then in the mid 19th c. it became the German Theatre; in 1945 it was renamed in honour of the Czech dramatist and actor Josef Kajetán Tyl (1808–56), and it is now the second house of the National Theatre. Restoration work on the theatre should be completed by the end of 1991.

The triumphant first performance of Mozart's "Don Giovanni" took place in this theatre on 29 October, 1787. From 1813 to 1816 Carl Maria von Weber (1786–1826) was musical director here.

Location
Železná ulice 11, Staré Mešto,
Praha 1

Metro
Můstek

*Týn Church (Kostel Panny Marie před Týnem) — E4

Location
Staroměstské náměstí (Old
Town Square), Staré Město,
Praha 1

Metro
Staroměstská, Můstek

Trams
5, 9, 19, 29

This Gothic church is the landmark and emblem of Prague's Old
Town. An aisled church with three presbyteries, it was built in
1365 on the site of an earlier Romanesque church; the choir was
completed in 1380; and the façade and the high-pitched roof
were built in the reign of George of Poděbrad (1460). At the time
of the Hussite Reform movement, the Týn Church was the principal church of the Utraquists of Bohemia (who believed that the
Communion should be administered in both kinds). Among
those who preached here were Konrad Waldhauser, Jan Milíč of
Kroměřiže and the Hussite Archbishop Jan Rokycana.

George of Poděbrad caused the façade to be adorned with a large
gilded chalice (an emblem of the Utraquist doctrine) and a statue
of himself; but after the defeat of the Protestants in the Battle of
the White Mountain (1620) the chalice was replaced by an image
of the Virgin.

Exterior

The 80 m (260 ft) high towers, the spires, of which are surrounded
by four elegant little turrets, were built in 1463–66 (the north
tower) and 1506–11 (the south tower).

After the great fire of 1679 the nave was given a new Baroque
vaulted roof. Note the fine north doorway with its decorated
Gothic canopy and tympanum from the workshop of Peter Parler
("Christ's Passion": a copy of the original, which is in the
National Gallery's Collection of Bohemian Art in Hradčany Castle
(see entry).

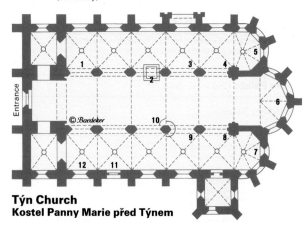

Týn Church
Kostel Panny Marie před Týnem

1 St Adalbert's Altar
2 Late-Gothic Baldachin
  St Luke's Altar
3 St Joseph's Altar
4 Altar of the Annunciation
5 Gothic Calvary group
6 High Altar ("Assumption", "Holy
  Trinity", by K. Škréta)

7 Gothic corbels pewter font
  (1414)
8 St Barbara's Altar
9 Marble tomb of the Astronomer
  Tycho Brahe (1601)
10 Renaissance altar
11 Gothic Madonna with Child Jesus
12 Altar of St Wenceslas

The interior, with the High Gothic choir and the large Baroque altars, is rather dark.

On the high altar is a fine "Assumption" and "Holy Trinity" by Karel Škréta (1610–74).

In the chapel to the left of the choir is a Gothic Pietà (15th c.). Gothic corbels from the Parler workshops adorn the end of the right aisle and there are also busts of unknown monarchs and a marble statue of St Cyril and St Methodius by Emanuel Max (1847). In addition there is a fine Gothic font made in 1414 from pewter. The altar-piece of St Adalbert to the left of the entrance is by K. Škréta who also did the paintings for the "Annunciation" altar and the St Barbara altar. The picture for the St Joseph altar was also painted by him (1664). A Late-Gothic baldachin (1495) by Master Matthias Rejsek of Prostějova hangs above the 19th c. neo-Gothic altar dedicated to St Lucas. Its altar-piece is the work of Josef Hellich.

On the fourth pier to the right of the main entrance can be seen the gravestone of the Danish astronomer Tycho Brahe (1546–1601: see Notable Personalities) who was Court Astronomer to Rudolf II. The two Latin mottoes next to his portrait read, "To exist rather than to shine" and "not power nor riches – only art is forever". There is a Gothic Madonna and Child in the right aisle (c. 1400). The relief of the Baptism of Christ was made for the Renaissance altar at the beginning of the 17th c. and A. Stevens painted the picture for the St Wenceslas altar at the end of the 17th century.

Týn Court (Týnský dvůr)   E4

This medieval trading centre (at present in course of renovation), also known as Ungelt (geld = "money") after the dues which had to be paid here, was established as early as the 11th c. Within this area, under the protection of the ruling prince – for which they paid a fee – merchants coming to Prague to do business stored, sold and paid customs dues on their wares up until 1773.

Location
Staré Město, Praha 1
Týnskýdvůr – Ungelt

Metro
Staroměstská, Můstek

Trams
5, 24, 29

The finest building within the complex is the Granovský Palace, a Renaissance mansion with an open loggia on the first floor (1560) in which visiting merchants could lodge. The wall-paintings in the loggia depict Biblical and mythological scenes. The doorway shows the year, 1560, and also bears the arms of the Granovský family.

Take note of the adjoining houses which were built between the 14th and 19th c.

Týn School (Týnska škola)   E4

Originally Gothic, with a rib-vaulted arcade, this building was enlarged in the middle of the 16th c., remodelled in the style of the Venetian Renaissance and given a double pediment.

Location
Staroměstské náměstí 14
(Old Town Square),
Staré Město, Praha 1

Metro
Staroměstská, Můstek
Náměstí Republiky

From the beginning of the 15th c., for over 400 years, this was the Týn parish school. The famous master-builder Matthias Rejsek of Prostějov taught here during the mid 15th c.

The Týn Church is through the third arch from the left.

Tyrš House (Tyršův dům) **and Museum of Physical Education and Sport** (Muzeum tělesné výchovy a sportu)

E3

The Tyrš House is occupied by the Museum of Physical Education and Sport, with a large collection of material illustrating the development of sport in Czechoslovakia. A special section is devoted to the Sokol gymnastic movement and the Spartakiades (a kind of national equivalent of the Olympic Games).

Location
Újezd 40, Malá Strana, Praha 1

Trams
12, 22

This little Renaissance palace was originally built about 1580. After the Emperor's victory in the Battle of the White Mountain (1620) it was acquired by Pavel Michna Vacinov, who had grown rich from the confiscated property of the rebellious Bohemian nobles, and his son had it remodelled in the style of the Late Renaissance and extended by the addition of an east wing.

Opening times
Museum: Tues.–Sat.
9 a.m.–5 p.m.
Sun. 10 a.m.–5 p.m.

From 1767 the mansion was used as an arsenal. After the First World War it was acquired by the Sokol movement and given its present name in honour of the founder of the movement, Miroslav Tyrš.

*Václav Vack Square (Náměstí primátora dr. Václava Vacka)

E4

This square was formerly known as St Mary's Square after a church dedicated to the Virgin which formerly stood here and played an important part during the German Reformation. As a result of building development in the early 20th c., however, only the south and west sides of the square remain in original form.

Location
Staré Město, Praha 1

Metro
Staroměstská

Municipal Library (Městská lidová knihovna)

Opening times
Tues.–Sun. 10 a.m.–5 p.m.

On the north side of the square stands the Municipal Library, opened in 1928. The balcony at the front has six allegorical statues by L. Kofranek. On the second floor of the library is the National Gallery's Collection of Modern Art (Sbírka moderního umění), mainly consisting of works by 20th c. Czech artists (V. Spála, J. Zrzavý, etc.). This collection is due to be transferred to the Museum for 20th c. Czech Art as soon as it moves into its new exhibition rooms in the reconstructed Trade Fair Pavilion (in Praha 7).

The library has over 750,000 volumes as well as a large music collection. In addition to the reading-rooms there are a puppet theatre and a film theatre.

New Town Hall (Nová radnice)

On the east side of the square is the New Town Hall (1909–12), in a late style of Jugendstil (Art Nouveau), with the office of the Primator (Burgomaster), the council chamber of the National Committee and the offices of the municipal administration.

At the ends of the façade (by L. Šaloun) are statues of the Iron Knight and of Rabbi Löw, who was reputed to have created a golem (man-made human being). The allegorical relief at the entrance to the Town Hall as well as the figures representing "Revision" and "Accountancy" are by S. Sucharda. The group of statues on the balcony, "Modesty", "Strength" and "Perseverence" are the work of J. Mařatka. At the south-east corner of the

◄ Týn Church

Václav Vack Square

Václav Vack Square: New Town Hall

Clam-Gallas Palace

square, against the wall of the Clam-Gallas Palace, is a fountain with a figure (by Václav Prachner, 1812) representing the Vltava, familiarly known to the people of Prague as Terezka.

Clam-Gallas Palace (Clam-Gallasův palác)

The Clam-Gallus Palace – entrance in Husova třída (Hus Street), which runs south from the square – now houses the Municipal Archives.

This magnificent Baroque palace, designed by Viennese architect Johann Bernhard Fischer von Erlach, was built in 1707 for Count Johann Wenzel Gallas. Stone giants guard the gate on Hus Street (Husova třída) which crosses Charles Street (part of the royal route). When the palace was rebuilt at the end of the 1980s it was somewhat of a surprise to discover that the heads of the three-metre-high figures had been replaced at the beginning of the century by artificial stone copies. The giants flanking the door-ways, the figures on the attic storey and the statue on the fountain in the first courtyard were the work of Matthias Braun. The frescoes on the staircase were painted by Carlo Carlone (1727–30), who was also responsible for the ceiling-paintings in two rooms on the second floor ("Olympus", "Coronation of Art and Learning") and in the library ("Luna, Helios and the Stars").

On the west side of the square is the extensive complex of the Clementinum (see entry).

The Villa Amerika, home of the Dvořák Museum

Valdštejn Palace, Valdštejn Street

See Waldstein Palace, Waldstein Street

Villa Amerika (Letohrádek Amerika) and Dvořák Museum   G5

Location
Ke Karlovu 20, Nové Město,
Praha 2

Metro I. P. Pavlova

Buses 148, 272

Trams 4, 6, 16, 22

Opening times
Tues.–Sun. 10 a.m.–5 p.m.

This Baroque mansion was built by Kilian Ignaz Dientzenhofer in 1717–20 as a summer residence for Count Michna (after whom it is also known as the Michna Palace). The richly patterned and finely articulated architecture of the main front make this one of the finest secular buildings of the Baroque period in Prague. A copy replaced the original Baroque wrought-iron work at the entrance. The frescoes in the interior are by J. F. Schor (1720). The statuary in the garden came from the workshop of Anton Braun. The Dvořák Museum now housed in the villa possesses scores and documents relating to the composer, in particular his correspondence with Hans von Bülow and Johannes Brahms.
Opening times: Tues.–Sun. 10 a.m.–5 p.m.

Vrtba Palace (Vrtbovský palác)   E3

Location
Karmelitská ulice 25, Malá
Strana, Praha 1

Trams
12, 22

Opening times
May–Sept. daily 8 a.m.–7 p.m.

The Vrtba Palace was rebuilt in Late Renaissance style in the 1630s. Its gardens (Vrtbovská zahrada), which were designed by F. M. Kaňka, are among the finest Baroque gardens in Central Europe.

The paintings in the *sala terrena* are by W. L. Reiner. At the entrance to the former vineyard are statues of Bacchus and Ceres by Matthias Braun (c. 1730). On the balustraded double staircase Baroque vases alternate with figures from Greek mythology. From the uppermost terrace there is a superb view of St Nicholas's Church (see Lesser Quarter Square) and the old Town.

Vyšehrad   H4

Location
Praha 1

Metro
Vyšehrad

Buses
134, 138, 148

Trams
3, 17, 21, 27

According to legend the crag of Vyšehrad was the spot where the Princess Libuše or Libussa stood and prophesied the future greatness of Prague and the site of the stronghold of the first Přemyslid rulers (see Quotations – Adalbert Stifter).

The Vyšehrad (High Castle) was probably founded in the 10th c. as the second castle of Prague. The first documentary reference to the castle, however, is in the reign of King Vratislav (1061–92), who transferred his residence from the Hradčany (see entry) to here. In those days the Hradčany was the seat of the bishop. Vratislav built a stone castle and several churches (SS Peter and Paul, St Lawrence's, etc.) on the rock above the Vltava and founded the collegiate chapter, long to be an important centre of culture. The Codex Vyssegradensis, now in the manuscript collection of the Clementinum (see entry), was produced here. The only surviving building of this period is the Round Chapel of St Martin. Soběslav I continued the building activity of his predecessors, but after his death in 1140 the Vyšehrad was neglected in favour of the Hradčany, to which the Bohemian kings now transferred their principal residence.

Charles IV carried out extensive renovation work, surrounding the castle with a circuit of walls which joined up with the town walls.

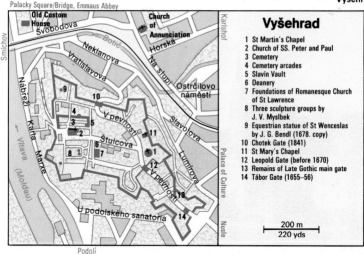

Palacky Square/Bridge, Emmaus Abbey

Vyšehrad

1 St Martin's Chapel
2 Church of SS. Peter and Paul
3 Cemetery
4 Cemetery arcades
5 Slavín Vault
6 Deanery
7 Foundations of Romanesque Church of St Lawrence
8 Three sculpture groups by J. V. Myslbek
9 Equestrian statue of St Wenceslas by J. G. Bendl (1678. copy)
10 Chotek Gate (1841)
11 St Mary's Chapel
12 Leopold Gate (before 1670)
13 Remains of Late Gothic main gate
14 Tábor Gate (1655–56)

200 m
220 yds

Podolí

The traditional coronation procession of the Bohemian kings started from the Vyšehrad and proceeded by way of Charles Square, Old Town Square and the Charles Bridge (see entries) to St Vitus's Cathedral in the Hradčany.

Coronation Procession

During the Hussite Wars, in 1420, almost all the buildings on the Vyšehrad were destroyed, and thereafter craftsmen and tradesmen established the "free town on the Vyšehrad". In the latter part of the 17th c. a Baroque fortress was built on the Vyšehrad and the burghers' houses were demolished. The fortress was dismantled in 1866 when the Vyšehrad became a "Quarter", and the cemetery was laid out. Finally in 1911 the fortress was razed to the ground, leaving only the circuit of walls.

The Vyšehrad, which has been declared a national monument, is at present in course of restoration. The legendary Vyšehrad was a favourite theme of graphic artists, musicians and writers, particularly during the Romantic period in the 19th c. Well-known works include Bedřich Smetana's opera, "Libusse", Felix Mendelssohn-Bartholdy's composition, "Libusse's Prophecy" and Franz Grillparzer's drama, "Libusse".

The best plan is to approach the Vyšehrad by Vratislav Street (Vratislavova) and enter by the Chotek Gate (1841) on the north side. To the right of the gate is a copy of J. G. Bendl's equestrian statue of St Wenceslas (1678).

The street called V pevnosti leads to the oldest building in Prague, the Romanesque Round Chapel of St Martin (Rotunda svatého Martina), which dates from the time of King Vratislav. When the Vyšehrad became a fortress in the 17th c. the chapel was used as a powder-magazine. It was renovated in 1878. The statues by J. V. Myslbek in the adjoining gardens represent figures from Czech legend. From the chapel Štulcova ulice leads to the Deanery, behind which are the foundations of the Romanesque Church of St Lawrence.

St Martin's Chapel

Vyšehrad: the Hochburg above the Vltava

Church of SS Peter and Paul

The twin towers (J. Mocker and F. Mikš) of the Church of SS Peter and Paul, which date only from 1902, have become the principal landmarks of the Vyšehrad. The church itself was built in the second half of the 11th c. It was rebuilt as an aisled Gothic basilica in the time of Charles IV and in the 16th c. was remodelled in Renaissance style. At the beginning of ther 18th c it underwent a Baroque transformation by F. M. Kaňka and C. Canevales, and between 1885 and 1887 was renovated in neo-Gothic style. Extensive restoration work was carried out between 1981 and 1987 in connection with archaeological excavation.

Notable features of the interior are an 11th c. stone sarcophagus in the first chapel on the right ("The tomb of St Longinus") and in the third chapel on the right a 14th c. panel-painting of the "Madonna of Rain" (invoked in time of drought), which is believed to have come from the Emperor Rudolf II's Collection. The main altar by J. Mocker, has statues of four saints (SS Peter and Paul, Cyril and Methodius) which were the work of F. Hrubes (latter part of the 19th c.). The frescoes with elegant plant decoration were painted in 1902–03 by V. Urbanova and F. Urban.

**Cemetery
(Vyšehradský hřbitov)**

Immediately north of SS Peter and Paul is the Vyšehrad Cemetery, a national shrine for distinguished representatives of art and culture which was created by the extension of the old medieval churchyard after the fortress was dismantled in 1866. Among those buried in the cemetery and the arcades surrounding it are the composers Smetana and Dvořák and the writers, Božena Nemcová, Krel Čapek and Jan Neruda as well as the artist Mikoláš Aleš. In the Slavín Vault (by Antonin Wiehl and J. Maudr) are buried the sculptors, J. V. Myslbek, Bohumil Kafka and Ladislav Šaloun, the artist, Alfons Mucha and violinist Jan Kubelík.

At the south-east end of the Vyšehrad (fine views) stand the Leopold Gate (Leopoldova brána) and, in a projecting outwork, the Early Baroque Tábor Gate (Táborská brána).

Gates

Waldstein Palace (Valdštejnský palác) **and Waldstein Gardens**   D3
(Zahrada Valdštejnského paláce)

The Waldstein Palace now houses the Ministry of Education and the Komenský Museum (Tues.–Sun. 9 a.m.–12 noon, 1–4 p.m.), which contains material on the pedagogue and philosopher better known as Comenius (1592–1670). This most sumptuous of Prague's noble residences and its first Baroque palace was built in 1624–30 for Albrecht von Waldstein (Wallenstein), one of the wealthiest nobles of his day, Imperial Generalissimo during the Thirty Years' War and later Duke of Friedland (murdered in 1634).

Waldstein had 25 houses, three gardens and one of the town gates destroyed in order that he might build his palace facing the Hradčany (see entry). The plans of the palace were drawn up by Andrea Spezza and Giovanni Pieroni, and the work was carried out under the direction of Giovanni Battista Marini. In the Hall of Knights in the front wing, there is a ceiling-painting portraying Albrecht von Waldstein as the god Mars in his triumphal carriage (B. Bianco, 1630). In the other rooms, paintings include a portrait of Waldstein on his horse (F. Leux, 1631) and one from the 19th c. by P. Maixner depicting classical motifs. The palace chapel has the oldest Baroque altar in Prague (E. Heidelberger).

Location
Malá Strana, Praha 1
Valdštejnské náměstí

Metro
Malostranská

Trams
12, 22

Opening times
Gardens and Sala Terrena:
May–Sept. 9 a.m.– 7 p.m.

Waldstein Palace

In the words of Golo Mann in his "Wallenstein":
"The front is Bohemian Italian, modelled on the Palazzo Farnese . . . The true dimensions of the palace can be grasped only after inspecting the inner courtyards and the park. From the square only the façade can be seen . . .

"The rest – the whole – was no ordinary palace. It was an independent territory, a miniature kingdom amid the huddle of the city, enclosed by subsidiary buildings and a fortress-like park wall. When Wallenstein's carriage rolled into the courtyard to the left of the main front he had everything he needed: a chapel to worship in; a riding-track at the lower end of the park; a bathing grotto with crystals, shells and stalactites; garden walks with statues and fountains."

Waldstein Gardens

The Waldstein Gardens, laid out in Italian Baroque style with grottoes, a pond and an aviary, can be entered from Letenská ulice. Along the walks and on the fountain are copies of bronze statues by the Dutch sculptor Adriaen de Vries, who was then working in Prague; the originals were carried off by the Swedes during the Thirty Years' War and are now at Drottningholm Palace, Stockholm. From the gardens there are attractive views of the Hradčany (see entry) and St Vitus's Cathedral. On the west side is the Sala Terrena, designed by Giovanni Pieroni, with frescoes by Baccio del Bianco. Theatrical performances and concerts are given here in summer.

On the north side of the palace is Waldstein Street (see entry).

Waldstein Street (Valdštejnská ulice)   D3

Location
Malá Strana, Praha 1

Metro
Malostranská

Trams
12, 22

Waldstein Street preserves to perfection the style and atmosphere of Baroque Prague. At No. 14, on the left, is the Palffy Palace. At No. 12 is the entrance to the Ledebour Gardens (Ledeburska zahrada), which extend into the gardens of the other palaces in the street and from the steep terraces, with loggias and a pavilion, afford impressive views over the city.

At No. 10 is the Kolovrat Palace (18th c.; now housing part of the Ministry of Culture) and at No. 8 the Fürstenberg Palace (1743–47), now the Polish Embassy (garden not open to the public). Diagonally across the street on the right-hand side can be found the former riding school of the Waldstein Palace (Jzdárna Valdštejnského paláce), now used for exhibitions.

Wenceslas Monument (Pomnik svatého Václava)   F5

Location
Václavské náměstí (Wenceslas Square), Nové Město, Praha 1

Metro
Muzeum

In front of the National Museum (see entry), at the south-east end of Wenceslas Square (see entry), is the Wenceslas Monument (by Josef Václav Myslbek, 1912–13).

Wenceslas ruled as Duke of Bohemia from 921 and was murdered by his brother Boleslav in 929 or 935. Following reports of miracles he was canonised and became the Patron Saint of Bohemia (feast day 28 September). Although his murder was in fact an

St Wenceslas, Bohemia's Patron Saint

incident in the struggle for power in Bohemia between Saxony and Bavaria, he was honoured as a martyr. Wenceslas (Václav) has long been one of the commonest boys' names in Bohemia. The journalist and traveller, Johann Fischart (1546–90) claimed that Bohemians are called Wenzel and Poles Stenzel (Stanislaus). The statue is surrounded by figures of four other patron saints of Bohemia. In front is St Ludmilla "Who is kind to the people", Wenceslas's grandmother and wife of the first Duke of Bohemia to be baptised. After her murder by pagan opponents she was recognised as Bohemia's first martyr.

Also in front of the statue, on the left, is St Procopius. To the rear are the Blessed Agnes (Anežka) and St Adalbert of Prague (Vojtěch).

There are other statues of St Wenceslas in Knights of the Cross Square (Vintners' Column) and the Vyšehrad (see entries).

---

*Wenceslas Square (Václavské náměstí)   E/F5

Wenceslas Square, 750 m (820 yd) long and 60 m (66 yd) wide, is more like a boulevard than a square. It is the centre of modern Prague, surrounded by hotels steeped in tradition such as the Europa, shops and offices, restaurants and cafés, cinemas and intimate theatres. With the neighbouring streets (Na Příkopé – see entry, Na Můstku, 28. řijna and Národní) it forms the "Golden Cross", where the commercial and social life of the city has developed most intensively down the centuries.

Location
Nové Město, Praha 1

Metro
Můstek, Muzeum

Trams
3, 9, 24

Wenceslas Square

During the construction of the Metro, Prague's Underground, in recent years the square has been replanned, with particular concern for the interests of pedestrians.

The square was originally laid out in the time of Charles IV as a horse market. It was given its present name in 1848. As an arena for political rallies, Wenceslas Square played an important role in the most recent demonstrations by the Civil Rights Movement which took place here at the end of the 1980s.

At the south-east end of the square is the National Museum, with the Wenceslas Monument (see entries) in front of it.

White Mountain (Bilá Hora) and Star Castle (Letohrádek Hvězda)

Location
Břevnov, Praha 6

Buses
108

Trams
8, 22

This bare limestone hill (318 m/1737 ft above sea-level) on the western outskirts of the city, now partly built up, was the scene of the Battle of the White Mountain on 8 November 1620 which decided the destinies of Bohemia under the Habsburgs. Here, in less than an hour, the Protestant nobility's army of mercenaries commanded by Count Matthias von Thun was defeated by the forces of the Catholic League under Maximilian of Bavaria. Elector Frederick of the Palatinate, elected as King of Bohemia by the Estates under a new constitution for an electoral monarchy, was compelled to flee from Prague (the "Winter King", 1619–20), and the country lost its independence, not to be recovered until 1918. A chapel was later erected on the battlefield which, in the 18th c., became the Church of Mary the Victorious. A little to the north is a memorial to the Battle of the White Mountain.

Star Castle on the White Mountain

Star Castle

On the north-western slopes of the hill, in a former zoological garden (Obora Hvězda), is Star Castle.

In 1530 King Ferdinand I established a game park in the Forest of Malejov which later came to be used for royal festivities and marksmanship contests. In 1797 the game park was laid out as an English-style park with broad promenades, called the Star Park (Obora Hvězda) after the old hunting-lodge of that name.

The castle lies on the north side of the park's main avenue. This unusual Renaissance building in the form of a six-pointed star, with a completely undecorated exterior, was erected by Italian architects (1555–58) as a hunting-lodge for Ferdinand of Tyrol and became the residence of his future wife Philippine Welser, daughter of an Augsburg patrician. It was later used as a powder-magazine. It now houses a museum devoted to the Czech writer Alois Jirásek and the painter Mikoláš Aleš. Concerts and literary recitals are given here occasionally. On the lower floor there is an exhibition which illustrates the historically important Battle of the White Mountain.

The glazed Renaissance tiles in the old refectory on the second floor are interesting.

The interior has charming Italian stucco decoration (1556–63), consisting of 334 ceiling-panels with scenes from Greek mythology and Greek and Roman history.

175

Zbraslav Castle with Collection of 19th and 20th c. Czech Sculpture

Location
10 km (6¼ miles) south of Prague

In the second half of the 13th c., Ottokar II had a hunting lodge and chapel built on the confluence of the Vltava and the Berounka which, during the reign of King Wenceslas II was converted into a Cistercian monastery. Rebuilding of the monastery, destroyed during the Hussite Wars, took place at the beginning of the 18th c. under the direction of G. Santini-Aichl and F. M. Kaňka. This was suspended in 1784 and at the beginning of the 20th c. the building was turned into a castle complex made up of three parts. Since 1976 the National Gallery's Collection of Czech Sculpture has been housed here and in the surrounding gardens. The sculpture on display is the work of 19th and 20th c. artists.

Exhibits worthy of note are: The Vltava Allegory (1812) by V. Prachner; the statue of Adam and Eve by Václav Levý (1849) and the portrait on porcelain of Bernhard Bolzanos (1849) by Arnošt Bruno Popp. Leading artists of the late 19th c. include Josef Václav Myslbek, an advocate of the Romantic style reflected, for example, in his "2nd Music Sketch" (1892–94). In 1891 Bohuslav Schnirch made the model for a pylon on the front of the National Theatre in Prague. The later work by Stanislav Sucharda is marked by the decorative detail of the "Youthful" style and Symbolism is reflected in František Bilek's works ("The Blind Girl", 1926). The most important impressionist artists are namely, Josef Mařatka (Portrait of Antonín Dvořák, 1906) and Bohumil Kafka ("Doe with Young", 1905).

Zoological Garden (Zoologická zahrada)   A3

Location
Troja, Praha 7

Bus
112

Opening times
April daily 7 a.m.–5 p.m.
May daily 7 a.m.–6 p.m.
June–Sept. daily 7 a.m.–7 p.m.
Oct.–Mar. 7 a.m.–4 p.m.

The Prague Zoo was established in 1931 on 45 hectares (111 acres) of natural country, a mingling of pasture, woodland, hills, rocks and ravines. It now contains more than 2000 animals, representing 600 different species of mammals, fishes and lower forms of life. The terrarium was re-opened in 1989 with a larger number of cobras, tortoises, rattlesnakes, rare crustaceans and pythons.

The zoo achieved a particular triumph in successfully breeding Przewalski's horse, a wild horse which is now extinct in natural conditions.

Hotel Forum opposite the Palace of Culture ▶

Practical Information

It is not always possible to give addresses and/or telephone numbers for all places listed in the Practical Information Section of these guides. This information is readily obtainable from hotel reception desks or from the local tourist office.

Airlines

British Airways
Štěpánská 63, Tel. 24 08 47/8

Czechoslovak Air Lines
(Československé Aerolinie, ČSA)
Revoluční 1, Kotva, Staré Město, Tel. 21 46
Airport: Ruzyně, Tel. 3 34

Lufthansa, Parizska 28, 11000 Praha 1, Tel. 2 31 75 51

Airport (Letiště)

Prague's international airport is 20 km (12½ miles) north-west of the city centre at Ruzyně. It can be reached by way of Dejvice (Leninova) or Břevnov (Belohorská).

Airport buses depart from and arrive at the ČSA Office, Revoluční 1 and the Interhotels Alcron, Ambassador, Esplanade, Forum, Intercontinental, Panorama, Paříž, Parkhotel and Zlatá Husa.

British Airways – Tel. 334 Ext. 4421.
Pan American – Tel. 26 67 47/8 (city).

Antiques (Starožitnosti)

Praha 1 (Nové Město), Václavske náměstí 60
Praha 1 (Staré Město), Uhelný trh 6
Praha 1 (Nové Město), Mikulandská 7
Praha 1 (Staré Město), Královdorská 2
Praha 1 (Staré Město), Můstek 384/3
Praha 2 (Vinohrady), Vinohradská 45
Praha 7 (Holešovice), Šimáčkova 17

Books

Praha 1 (Staré Město), Dažděná 5
Praha 1 (Malá Strana), Mostecká 22
Praha 1 (Staré Město), Karlova 2
Praha 1 (Staré Město), Ulice 28. řijna 13
Praha 2 (Nóve Město), Ječna 36

Coins

Praha 1 (Staré Město), Melantrichova 9

Pictures and graphic art

Praha 1 (Staré Město), Karlova 14

Art Galleries

Collection of Graphic Art
See A to Z – Palais Kinsky

National Gallery Collections

Collection of Old Bohemian Art in St George's Convent
See A to Z – Hradčany

Collection of European Art in the Palais Sternberg
See A to Z – Collection of European Art

Collection of Modern Art in the City Library
See A to Z – St Mary's Square

Collection of Bohemian Applied Art and 19th century Czech
Painting
See A to Z – St Agnes' Convent

Collection of Czech Sculpture of the 19th and 20th centuries
See A to Z – Zbraslav Castle

František-Bilek-Villa
Praha 6 (Dejvice), Mickiewiczova 1
Daily (except Monday) 10 a.m.–noon, 1–6 p.m.

Other galleries and
exhibition rooms (selection)

Fotochema
Praha 1 (Nové Mešto), Jungmannovo Náměstí 16
Daily (except Monday) 10 a.m.–noon, 1–6 p.m.

Galerie D
Praha 5 (Smichov), Matoušova 9
See A to Z – Dienzenhofer Summer Residence
Daily (except Monday) 10 a.m.–1 p.m., 2–6 p.m.

Gallery of the Čapek Brothers
Praha 2 (Vinhorady), Jugoslávská 20
Daily (except Monday) 10 a.m.–1 p.m., 2–6 p.m.

Prague City Gallery
Exhibition hall in the Old Town Hall
See A to Z – Town Hall of the Old Town

Galerie 5
Praha 1 (Nové Město), Vodičkova 10
Daily (except Monday) 10 a.m.–1 p.m., 2–6 p.m.

Palace of Culture (Paláce kultury)
See A to Z – Palace of Culture

Mánes Exhibition Hall, "Gallery of the Young"
See A to Z – Mánes Exhibition Hall

Gallery of Central Bohemia
Varied collections
Praha 1 (Staré Město), Husova ulice 19
Daily (except Monday) 10 a.m.–1 p.m., 2–6 p.m.

New Hall (Nová siň)
Praha 1 (Nové Město), Voršilská 3
Daily (except Monday) 10 a.m.–1 p.m., 2–6 p.m.

Palace of the Nobility of Kunštat and Poděbrady
See A to Z – Podiebrad Palace

Riding School of Waldstein Palace
Praha 1 (Mallá Strana), Valdštejnská 2
See A to Z – Waldstein Street

Václav-Spála-Galerie
Praha 1 (Nové Město), Národni 30
Daily (except Monday) 10 a.m.–1 p.m., 2–6 p.m.

Banks (Banku)

Czechoslovak State Bank

The country's principal bank is the Československá Státní Banka
(Czechoslovak State Bank).
Praha 1, Na Příkopě 28.
Branches in district (county) towns throughout Czechoslovakia.

Opening times

In the centre of Prague Mon.–Fri. 8 a.m.–5 p.m., Sat. 8.30 a.m.–
1 p.m.
Other banks normally Mon.– Fri. 8 a.m.–noon or 2 p.m.

Exchange offices

Most banks and the most important travel agencies (Čedok, Pra-
gotur, Autoturist) have an exchange desk where money can be
changed. At branches of the State Bank foreign currency can be
used to buy Tuzex vouchers (see Currency).

Illicit money-changing

Almost all visitors are approached, in their hotel or in the street,
by people offering to change money. Deals of this kind are prohibited by law and may attract severe penalties.

Beer (pivo)

General

In Prague it is not uncommon, especially at midday, to see some-
one carrying a small tray going into the nearest beer-saloon, only
to re-emerge with six freshly filled tankards and hurry back to the
office. Beer is the national drink of Czechoslovakia, and in Prague
there are countless opportunities to sample the products of the
national breweries. Prague landlords look after their beer; many
are able to store in cellars with Gothic arches, where it is kept
exceptionally fresh, before being dispensed through taps which
are seldom idle.

Czech beer owes its world-wide fame principally to the excellent
hops which are grown in northern Bohemia around Saaz, Roud-
nitz, Auscha and Dauba, and which are exported all over the
world. There are records of hop-growing in Bohemia in 9th-
century chronicles, and from the 12th century onwards hops
were exported along the Elbe to the famous hop-market in Ham-
burg. For centuries the export of hop-plants was a capital offence.
Beers, which were renowned even in late medieval times for their
variety and quality, are brewed from barley, wheat and oats with
the particularly soft water of the country, and flavoured with the
excellent hops.

Types of beer

Czech beers are mostly bottom-fermented. Fundamentally they
are divided into light (světlé) and dark (tmavé). The details on
labels, bottle-caps, etc, do not refer to the alcoholic content of the
beer. The gravity is given in "grad plato", that is the proportion of
soluble material in the wort before fermentation. The alcoholic

content by weight is approximately one-third of the given gravity. Beers are also classed as "draught" (up to 10°; 3-4% alcohol), "lager" (up to 12°; about 5% alcohol) and special beers (13° upwards).

The best-known Bohemian beer is Pilsner Urquell (Plzeňský Praz-droj) from Pilsen, which has been brewed since 1842 and which is the model for all beers of the same type throughout the world. The first bar in Prague to serve Pilsner was the U Pinkasů ("Pin-kas's"; Praha 1, Jungmannovo náměstí 150, founded by a tailor named Pinkas in 1843 and which still exists today. However, the best Pilsner is served at U zlatého tygra ("The Golden Tiger"; Praha 1, Husova 17). Other well-known Pilsner bars, where good food is provided as well as beer are: U kocoura ("The Cat"; Praha 1, Nerudova 2), U Schnellů ("Schnell's"; Praha 1, Tomášská 2) and U dvou koček ("The Two Cats"; Praha 1, Uhelny trh 10).

Pilsner

Undeservedly somewhat overshadowed by the beers of Pilsen is the lighter and sweetish Budweiser (Budvar) which has been brewed in the south Bohemian town since 1531. To sample this beer in Prague it is best to go to U medvikü ("The Little Bear"; Praha 1, Na Perštýne 7), an inn which in the Middle Ages was actually a brewery.

Budweiser

From the Großpopowitzer brewery in central Bohemia come two beers; one a finely hopped light lager (Velkopopoviký kozel) with a creamy head; the other a 14° dark special beer. In Prague these two beers are particularly well-kept at U černého vola ("The Black Ox"; Praha 1, Loretánské náměstí 1).

Großpopowitzer

There are still four breweries in Prague itself. The largest of the four and also the largest in the whole country is the brewery in the district of Smichover which produces a pale lager called Staropramen ("old spring"), Smichov beer can be found at U Glaubicü (Praha 1, Malostranské náměští 5) and at dvon srdú ("The Two Hearts"; Praha 1, U lužikchého semináře 38).

Prague breweries

Smichover

Two beers are brewed in the district of Braniker: a light 14° and a dark 12°. Both are served at U svatého Tomáše ("St Thomas's"; Praha 1, Letenská 12). This was once the cellar of the former Augustinian brewery which continued to produce its own beer until 1950. At U supa ("The Vulture"; Praha 1, Celetná 22), an inn which has existed since the 14th century, only the light beer is now served.

Braniker

The last remaining authentic domestic brewery in Prague is a veritable institution. This is U Flekü ("Fleck's"; Praha 1, Křemencova 11) where a strong dark 13° beer has been brewed since 1499. Although – like St Thomas's cellar – it is much frequented by tourists, nevertheless Fleck's has a unique atmosphere in its extensive smoky bars, and, in summer, in its beer-garden. The inn is also noted for its traditional Czech cabarets.

*U Flekü

The "regular" of the Czech author Jaroslav Hašek (see Notable Personalities) and also of Schwejk, the hero of his novels, was U Kalicha ("The Goblet"; Praha 2; Nabogišta 13), now, however, generally crowded with tourists.

U Kalicha

The Holešovice brews a 10° draught beer called Pražanka.

Holešovice

Pleasure boats on the Vltava

Boat trips on the Vltava

The departure point for passenger boats on the Vltava is at the Palacký Bridge (Palackého most). During the summer months there are regular trips which enable the visitor to get a good general impression of the city. From July until September the old-time paddle wheel steamer, the "Vyšehrad", leaves on three-hour evening river cruises at about 7 p.m. on Tuesdays, Thursdays and Fridays; a Bohemian wind-band plays light music and Bohemian refreshments and Prague beer are served. There are also services to various attractive places in the immediate surroundings of Prague, such as Zbraslav or the beautiful Slapy Lake, formed by a dam on the Vltava.

Detailed information can be had from the Čedok Tourist Office (see Information).

Cafés

Alfa, 1 (Nové Město), Václaské náměstí 15
Arco, 1 (Nové Město), Hybernská 16
City, 1 (Nové Město), Vodičkova 38
Columbia, 1 (Staré Město), Staroměstské náměstí 15
Evropa, 1 (Wenceslas Square), Václavské náměstí 29
Jahodovy (U Myšáků), 1 (Nové Město), Vodičkova 31
Jalta, 1 (Wenceslas Square), Václavské náměstí 45

Café Slavia

Kajetánka, 1 (Hradčany Square), Kajetánské zahrady
Malostranská kavárna, 1 (Malá Strana), Malostranské
náměstí 28
Obecnídům, 1 (Staré Město), Náměstí Republiky
Praha, 1 (Wenceslas Square), Václavské náměstí 10
Slavia, 1 (Staré Město), Národní třída 1
U zlatého hada, 1 (Staré Město) Karlova ulice 18
Vyšehrad, 2 (Vyšehrad), Palace of Culture (Palác kultury)

Camping sites

Camping sites are run by the Autoturist organisation, Opletalova
29. Places can be reserved by telephone (22 35 44–9).

Management

Caravancamp
5, Motoi, Plzeňska; tel. 52 47 14

Campsites
Category A

Na Vlachovce
Kobylisy, Rudé armády 217, Tel. 84 12 90
4 km (2½ miles) north of the city centre

Sokoi Troja
7, Troja, Trojska 171 a; Tel. 84 28 33

Dolni Chabry
8, Dolni Chabry, Ústecká ulice (mid-June to mid-Sept.)
12 km (7½ miles) north of the city centre

Category B

Car rental

PRAGOCAR (also agent for international car rental firms such as Europcar, Inter-Rent, Godfrey Davis, Avis, etc.):

Praha 1, Štěpánská 42, Tel. 235 28 25, 235 28 09
Ruzyně Airport, Tel. 36 78 07
Inter-Continental Hotel, Tel. 231 95 95

Chemists/Pharmacies (Lekarnu)

After-hours service

Praha 1 (Staré Město), Na Příkopě 7, Tel. 22 00 81
Praha 2 (Nové Město), Ječná 1, Tel. 26 71 81
Praha 6 (Břevnov), Pod Marjánkou 12, Tel. 35 09 67
Praha 10 (Vršovice), Moskevská 41, Tel. 72 44 76

Currency

Import and export of currency

Czechoslovak currency and the currencies of other Eastern Bloc States may not be brought into or taken out of Czechoslovakia. Other currencies may be imported or exported without limitation and without any declaration.

Czechoslovak currency

The unit of currency is the Czechoslovakian crown (koruna, abbreviated Kčs), which is divided into 100 heller (haléř, abbreviated hal). There are banknotes for 10, 20, 50, 100, 500 and 1000 crowns, and coins in denominations of 5, 10, 20 and 50 heller. The new 1000-crown notes are blue.

Exchange rate

Foreign tourists to Czechoslovakia are granted a "tourist rate of exchange". Tuzex vouchrs at the official rate of exchange can be obtained from Alimex, at exchange counters of the state bank (see Banks), or at Tuzex (see Shopping). These are necessary, in addition to hard currency, to make purchases in state Tuzex shops.
Currency exchanges at the "tourist rate" can only be effected at the exchange counters of the state bank, at Inter-hotels (see Hotels), at principal tourist offices or at frontier crossing posts. Inter-hotels and tourist offices generally give a better "tourist exchange rate".

Visitors are particularly warned against illegal exchange of currency.

Compulsory currency exchange

A certain amount of foreign currency must be exchanged as a condition of the granting of a visa and must be effected at the latest at the frontier. At present the compulsory amount for adults is a minimum of £10. Since 1989 children and young persons up to the age of 17 are exempted from this regulation, as are tourists who have booked a package deal, holders of credit cards which are valid in Czechoslovakia (details in the visa), officially invited guests (details from the embassy) and visitors holding a transit visa valid for a maximum of 48 hours. The compulsory exchange amount is halved for those visiting relatives in the country. When leaving Czechoslovakia only the amount of currency noted in the visa exceeding the compulsory exchange can be reconverted.

The following credit cards are accepted: Access, American Express, Carte Blanche, Diners' Club, Eurocard, Master Card, Bank Americards, Visa and JCD.

Customs regulations

Personal effects, 3 kg of food, 250 cigarettes or the equivalent in other forms of tobacco, 2 litres of wine, 1 litre of spirits, and gifts to the value of 1000 Kčs may be imported into Czechoslovakia free of duty. All valuables (cameras, video equipment, calculators, etc.) must be entered in the valuation certificate (third side of the visa) which must be produced when leaving the country. No personal items may be offered for sale or given as presents without payment.

Arrival

Souvenirs to the value of 1000 Kčs may be taken out of the country without payment of duty. Special regulations and in some cases duty of up to 100% of the purchase price apply to such things as crystal, children's articles and antiques. Everything which has been purchased for foreign currency in TUZEX and ARTIA shops may be exported without payment of duty. Tuzex shops (see Shopping) will handle the formalities and arrange despatch.

Departure

It is advisable to keep receipts for petrol (particularly diesel fuel) until leaving the country.

Petrol station receipts

Department stores

Prior Kotva, Praha 1, náměstí Republiky 8
This store, which is always crowded, is the largest and most moden in Czechoslovakia. The complex was built in 1975 to plans by Věra and Vladimir Machonin of the Swedish firm of Saab. About a third of the total floor space of 60,000 sq. m (50,160 sq. yd) is used for sales. In addition to six shop floors there are nine levels of car-parking which is also available outside business hours. Among the goods on sale are textiles, food, housewares and office necessities. There are also a restaurant and a snack-bar.

Bílá labut' (White Swan), Praha 1, Na poříčí 23
A general department store with a range of goods second only to Kotva.

Prior Maj, Praha 1, Národni 26
Clothing, foodstuffs, industrial products.

Prior Dětsky dům, Praha 1, Na Příkopě 15
Children's store.

Družba, Praha 1, Václavské náměstí 21
China, leather goods, clothing, furniture, jewellery. Restaurant and café with outlook terrace.

Dům potravin, Praha 1, Václavské náměstí 59
Foodstuffs.

Kotva department store

Dům obuvi, Praha 1, Václavské náměstí 6
Footwear.

Dům mody, Praha 1, Václavské náměstí 58
"Fashion House"

Dům kožešin, Praha 1, Železná 14
Furs

Dům sportu, Praha 1, Jungmannova 28
Sports goods

Dům hudebních nástrojů, Praha 1, Jungmannovo náměstí 17
Musical instruments.

Electricity

220 volts AC; in some of the older parts of the city still 120 volts AC. A universal adaptor should be taken for electric razors, etc.

Embassies

United Kingdom   Thunovská 14, Praha 1,
                      Tel. 53 33 47–9, 53 33 40, 53 33 70

United States   Tržiště 15, Praha 1,
                      Tel. 53 66 41–8

Mickiewiczova 6, Praha 6,   Canada
Tel. 32 69 41

Emergencies

Dial 155   Doctor on duty

Dial 333   Ambulance

Dial 158   Police

Events

Matthias Fair in Julius Fučík Park of Culture and Recreation   February

Intercamera (International Exhibition of Audio-Visual   April
Technology)

Prague Spring Festival (Music festival with Czech and foreign   May–June
orchestras and soloists; primarily classical concerts)

Peace Run (international cycle run from Prague to Berlin and
Warsaw)

Concertino Praga (concerts with young performers). Three-day   June
festival of Czech wind bands at the beginning of the month in
Kolin.
International television festival "Golden Prague". Quadriennial
of theatrical technology (next in 1991).
Memorial ceremonies in Lidice

Prague Summer Programme with concerts, theatrical perfor-   July/August
mances, folk-groups and cultural events.

Spartaklade in Strahov stadium (every five years, next in 1995) August

International Jazz Festival   October

Excursions

There are many places in the vicinity of Prague which, with their   General map page 188
fine Bohemian fortresses and castles, make excellent venues for
excursions. These include Karlštejn, Konopiště, Česty Šternberk,
Křivoklát and Kokořín which can all be found in the A–Z section
under the appropriate heading. Beautiful woods and impressive
stalactitic caves in the Bohemian karst region await exploration
by lovers of nature. The memorial places of Lidice and Terezín
have particular historical significance, and these are both
described in the A–Z section.

Sightseeing tours are regularly organised by Čedok (see Informa-   Organised excursions
tion). Whole-day excursions can be made to the spa town of

Dresden
Děčín (Tetschen)

Liberec
(Reichenberg)

Krkonoše
(Riesengebirge)

Teplice
(Tep-
litz-
Schönau)

Svádov

Česká Lípa
(Böhm. Leipa)

Zákupy

Osečná

Hodkovice

ÚSTÍ nad Labem
(Aussig)

15

Úštěk
(Auscha)

Jestřebí

Kuřivody

Turnov
(Turnau)

835 E55

Bílina
(Bilin)

Litoměřice
(Leitmeritz)

Doksy
605

Bělá

Mnichovo
Hradiště

Kost 35

Mladá Boleslav
(Jungbunzlau)

Sobotka

Lovosice
(Lobositz)

Terezín

Vrutice

Kokořín

Mšeno

Doksany

Libochovice

Roudnice
(Raudnitz)

Liběchov

Vtelno

Benátky

Vlkava

Děřenice 32

Mělník

Nová Ves

Veltrusy

Dymokury

Dlouhopolsko

Třebíz

Slaný
(Schlan)

Kralupy

Líbeznice

Brandýs

Nymburk

Poděbrady

Nové
Strašecí

Brandýsek

Lysá

KLADNO

Mochov

Sadská

Lidice

Úvaly

Český Brod

Kolín

Unhošť

**Krivo-
klát**
(Pürglitz)

Beroun

Zbraslav

PRAHA
(PRAGUE)

Říčany

Kostelec

Bečváry

Sedlec

Kutná Hora
(Kuttenberg)

Koněprusy

Karlštejn

Jesenice

Mirošovice

Zdice

Lochovice

Řevnice

Pyšely

Sázava

Uhlířské
Janovice

Hořovice

Štěchovice

Zbraslavice

Dobříš

Slapy

Konopiště

Benešov
(Beneschau)

Nový
Knín

Neveklov

Ledeč

Příbram

Obory

Vlašim

Dolní
Kralovice

Milín

Sedlčany

Votice

Loužovice
pod Blanicem

Čechtice

Rožmital

Sedlec

Malá
Vožice

Košetice

Březnice

Zalužany

Petrovice

Pacov

Červ.
Řečice

Blatna

Orlík
(Worlik)

Staré
Sedlo

Zvíkov
(Klingenberg)

Podhradí

Milevsko
(Mühlhausen)

Oltyně

Tábor

Obratáň

Pelhřimov
(Pilgram)

Mirotice

Hor. Záhoří

Bernartice

Sudoměřice

Černovice

Tučapy

Kamenice

Drhovle

Písek

Albrechtice

Soběslav

Strakonice
(Strakonitz)

Šumava (Böhmerwald)
Passau

České Budějovice
(Budweis)

České Budějovice (Budweis)
Linz

Jindřichův Hradec
(Neuhaus)

Touristische
Höhepunkte

0 5 10 15 km

© Baedeker

Karlovy Vary with an additional visit to Lidice; to the famous beer-town of Plzň (Pilzen); to the castles of Karlštejn and Konopiště with a stop at the Slapy reservoir; to the south Bohemian town of Tabor with a visit to the neo-Gothic castle of Hluboká; to Kutná Hora and the castle of Česky Šternberk (both pearls of Bohemian Gothic); to central Bohemia and the collection of weapons in Orlík Castle which is not far from the reservoir of the same name and continuing to the fortress of Zvíkov; or to the Bohemian granite collection together with a visit to the castle of Kost, Half-day excursions include, for example, visits to the castles of Zbraslav and Konopiště combined with a visit to the Slapy reservoir, or a trip to Mělník, the centre of viniculture where the produce of the district can be tasted.

Visitors who are interested in traditional Czech architecture should visit one of the open-air museums (Skansen). The nearest, in Přerov and Labern (25 km/16 miles east), have typical buildings of the upper Elbe which are explained in a thematic exhibition. Several old buildings from the plain of the River Želivka, which was flooded when the Žvihov dam was completed in 1976 (drinking water supply for Prague), were moved to Kouřim, 40 km/25 miles west of the city.

Open-air museums

In the fertile land of the Elbe valley to the east of Prague lies the industrial town of Kolin (altitude 225 m/738 ft); the old part of which is dominated by the church of St Bartholomew. This 13th c. church in Transitional style has a Late-Gothic choir built by Peter Parler between 1360 and 1378. Although some of the houses in the town square and the adjoining streets have Baroque gables, the heart of their architecture is also Gothic. At the end of the 19th c. the bandleader František Kmoch (d. 1912) was working in Kolin and it was he who brought international fame to Bohemian wind music. The town played an important role in 1757, when about 7 km/4 miles west of the municipal boundaries the Prussian troops of Frederic II came face to face with the Austrian army of Maria Theresa.

Kolín

About 20 km/12 miles north-west of the international airport of Prague-Ruzyně lies Kladno (altitude 3384 m/1260 ft). The town received a boost in the mid 19th c., when coal was discovered in the vicinity. The story of the mining and iron and steel industry of Kladno is told in exhibits in the Baroque castle museum, which since 1895 has also housed an art gallery with temporary exhibitions.

Kladno

The Baroque castle of Veltrusy dates from the first half of the 18th c. Its fine rooms house valuable collections, especially of Chinese and Japanese porcelain and of fine English crystal chandeliers. Among other exhibits is a large 16th c. Brussels tapestry. In the extensive park, which was laid out in the English style, are some rare old trees. It was here in 1754 that the first trade-fair in Europe was held.
Nelahozeves, a short distance from Veltrusy, is the birthplace of the famous composer Antonin Dvořák (see Famous People). The house in which he was born has been furnished as a museum.

Veltrusy

Food and drink

The favourite meat is pork which is prepared in many different ways. Perhaps the national dish can be defined as pork with

Meat dishes

cabbage and dumplings, but on festive occasions pork may be replaced by a succulent roasted goose or duck. Game dishes such as haunch of venison or larded hare are excellent. Other specialities include steamed beef, boiled pork, roast loin of pork and various kinds of sausage. Prague ham is world-famous.

Fish

Fish dishes do not figure prominently on the menu. However, few families go without the traditional carp at Christmas. It is usually fried, but can be roasted, boiled or served in aspic.

Sauces

Sauces play a large part in all main meals; often a sauce consists simply of seasoned meat juices which may be reduced and cream added. A white sauce, flavoured with caraway or marjoram, is frequently served with both meat and vegetables.

Dumplings

Dumplings are an essential feature of Bohemian cuisine. They are prepared in a great variety of ways; as well as noodles, potato and bread dumplings are the usual accompaniments to main dishes. Bacon dumplings, filled with cabbage or spinach and roasted onions, are readily accepted as a main dish. The highlight of Bohemian speciality dumplings is without doubt the fruit dumplings which are prepared from a yeast dough. In his novel "Barbara oder die Frömmigkeit" (Barbara or Devoutness) Franz Werfel (see Notable Personalities) has devoted more than 1000 words to this delicacy. The dumplings are filled with cherries, apricots, apples, bilberries or most often plums. They are spread with sieved hard curd cheese or with poppy seed, and melted butter is poured over them.

Desserts

Many and varied are the desserts. As well as apple-strudel, little cakes and doughnuts are exceptionally tempting, but the pancakes beat all of them. Filled with curd cheese, jam or chocolate, they are not so thin as the crêpes of Brittany, but they are just as delicious.

A brief guide to the menu

chléb	bread
hovězi pečeně	roast beef
husa s knedlíkem a zelím	goose with dumplings and cabbage
husí drůbky s rýží	goose giblets with rice
jablkový koláč	apple cake
jablkový závin	apple strudel
jeleni	roast hare
kachna se zelím a knedlíkem	duck with cabbage and bread dumplings
kapr na modro s bramborem	carp au bleu with boiled potatoes
knedlíky ovecné	fruit dumpling
koroptev s červeným zelím	partridge with red cabbage
merunkové knedlíky	apricot dumplings
nolky	roast potatoes
olomoucké syrečky	Olomouc sour milk cheese
ovar s křenem a chlebem	boiled pork with horse-radish and bread

palačinky s tvarohem	pancakes with curd cheese
pečená štika	roast pike
pečené kuře s bramborovou kaší	roast chicken with potato purée
polévka gulášova	goulash soup
polévka slepiči	beef soup
pražský řízek s chřestem a bramborem	Prague schnitzel with asparagus and potatoes
pstruh na másle	trout in melted butter
ryba	fish
smažený kapr s míchaným salátem	fried carp with mixed salad
špekové/chlupaté/knedlílky se zelím	bacon dumplings with cabbage
srnčí kýta s kroketami a brusinkami	haunch of venison with croquettes and cranberries
šunka s oblohou	Prague ham, garnished
svestkové knedlíky s tvarohem a máslem	plum dumplings with curd cheese and butter
svíčková na smetaně	roast loin of beef with cream sauce
uzené s okurkou a chlebem	smoked meat with gherkins and bread
vdolky, lívance	pancakes
vepřové se zelím	roast pork with sauerkraut
vídeňský řízek s bramborovym salátem	Wiener schnitzel with potato salad
zajačí hřbet na smetaně	hare in cream sauce

See separate entry — **Beer**

The wine-houses of Prague sell some of the internationally known wines, but the most popular are the Czech wines including Melnik Ludmila, Žernosecke, Velkopavlické, Valtické and Primatorské. Wines from southern Moravia are also of very good quality, especially those from Znaim, Nikolsberg and Feldsberg, as are the Slovakian wines from the surroundings of Bratislava. — **Wine**

"Hard" spirits of Czech origin are slivovice (plum brandy), meruňkovice (apricot brandy), žítná or režná (corn schnapps) and jalovcová or borovička (types of gin). To cure a stomach disorder the Becherovka of Karlovy Vary (Karlsbad), a herb liquor, is recommended. — **Spirits**

Coffee is normally Turkish coffee, boiled and served with the grounds. Italian-style espresso coffee can be had, but must be specially asked for. Viennese coffee, with half a glassful of whipped cream, can also be obtained.
Also available are tea (čaj), milk (mléko) and fruit juice (orange, grapefruit). — **Soft drinks**

Getting to Prague

By road

The distance from the Channel to Prague is between 965 and 1130 km (600 and 700 miles). Perhaps the best route is to follow E 5, which runs from Ostend to Nuremberg, and then switch to E 12, crossing into Czechoslovakia at Waidhaus/Rozvadov.

Other frontier crossings are:

from West Germany	Schirnding/Pomezí
	Furth im Wald/Folmava
	Bayerisch Eisenstein/Železná Ruda
	Philippsreuth/Strážny
from Austria	Wullowitz/Dolní Dvořiště
	Gmünd/České Velenice
	Haugsdorf/Hatě
	Drasenhofen/Mikulov
	Berg/Petržalka
	Grametten/Nová Bystřice
	Laa an der Thaya/Hevlin
	Weigetschlag/Stúdanky
	Neu-Hagelberg/Halámky
	Hainburg/Bratislava

The distance to Prague from the German border ranges between 160 and 200 km (100 and 125 miles), from the Austrian border between 170 and 400 km (105 and 250 miles). On all these roads there is likely to be much heavy goods traffic. Most of the Czech motorways are still under construction, and only certain stretches are open to traffic.

By bus

There are regular services by Czechoslovak buses from Frankfurt, Munich and Vienna to Prague, arriving at Florenc bus station on Vítězného Února.

By rail

From Britain the most convenient services are via Paris from where there is a daily through train to Prague or via Cologne from where through carriages convey passengers to the Czechoslovakian capital. There are also good services to Prague from Frankfurt am Main, Nuremberg and Stuttgart, as well as from Zürich (with a change), Berlin and Vienna. From Frankfurt the journey takes 11 hours, from Zürich 14–18 hours, from Vienna 8 hours. Seat reservation is advisable. There are some night services with sleepers.

Trains arrive at Prague's Central Station at Ulice Vítězného února 16, which since its reconstruction in 1970 has had one of the most modern passenger concourses in Europe. The Hlavní nádraží Metro Station is at this station.

By air

There are direct flights from London to Prague by British Airways and Czechoslovakian State Airlines (ČSA). There are weekly direct flights from New York to Prague by ČSA, and Lufthansa has frequent services via Frankfurt.

Prague Airport at Ruzyně is 20 km (12½ miles) north-west of the city on the road to Kladno. There is a bus service from the airport to the city centre and the journey takes about an hour. There are also taxis and car rental facilities.

Czechoslovak Air Lines can arrange accommodation in Prague.

Hotels

Advance reservation of rooms is necessary.   Reservations

Hotels are divided into five categories: A* de luxe, A*, B*, B   Categories
and C.

A* de luxe, A* and some B* hotels are classed as Interhotels. In
these hotels visitors can look for excellent accommodation (private bath or shower; suites) and staff who can speak their language. They usually have exchange offices, souvenir shops, bars
and several restaurants.

Category B* can be classed as very good, category B as good.

Category C applies to more modest hotels. These hotels are not
handled by Čedok, but rooms can be booked either through the
ČKM-SSM Youth Travel Agency or directly at reception.

Alcron, Praha 1, Štěpánská 40, Tel. 2 35 92 16–24   Category A* de luxe
Esplanade, Praha 1, Washingtonova 19, Tel. 22 25 52
Intercontinental, Praha 1, náměstí Curieových, Tel. 2 31 18 12
Jalta, Praha 1, Václavské náměstí 45, Tel. 26 55 41–9

Ambassador, Praha 1, Václavské náměstí 5, Tel. 22 13 51   Category A*
Forum, Praha 4, Kongresová 140, Tel. 42 21 11, 41 01 11

Intercontinental Hotel

193

Hotels

Grand Hotel Europa . . .

. . . with Art-nouveau restaurant

International, Praha 6, náměstí Družby 1801/1, Tel. 32 10 51
Olympik I, Praha 8, Invalidovna, Sokolovska tr. 138, Tel. 82 87 41
Palace, Praha 1, Panská 12, Tel. 26 83 41
Panorama, Praha 4, Milevská 7, Tel. 41 68 58
Park, Praha 7, Veletržni 20, Tel. 38 07 01 11

Club Motel, Průhonice, in Průhonice (15 km/10 miles),
Tel. 72 32 41–9
Motel Kohopiště, in Benešov (45 km/28 miles), Tel. 20 53

Category B*

Admiral (Botel), Praha 5, Hořejší nábřeží, Tel. 54 74 45
Ametyst, Praha 2, Makarenkova 11, Tel. 25 92 56–59
Axa, Praha 1, Na poříčí 40, Tel. 24 95 57
Beránek, Praha 2, Bělehradská 110, Tel. 25 45 44
Central, Praha 1, Rybná 8, Tel. 2 31 92 84
Centrum, Praha 1, Na poříčí 31, Tel. 2 31 01 35
Cružba, Praha 1, Václavské náměstí 16, Tel. 24 06 07
Evropa (Interhotel), Praha 1, Václavské náměstí 25, Tel. 26 52 74
Flora, Praha 3, Vinohradská 121, Tel. 27 42 41
Olympik II (Interhotel; no rest.), Praha 8, Invalidovna, Tel. 83 47 41
Paříž (Interhotel), Praha 1, U Obecního domu 1, Tel. 2 32 20 51
Splendid (Interhotel), Overnecká 33, Tel. 37 54 51
Tatran, Praha 1, Václavské náměstí 22, Tel. 2 35 28 88
Víkov, Praha 3, Koněvova 114, Tel. 27 93 41
Zlatá Husa (Interhotel), Praha 1, Vaclavské náměstí 7,
Tel. 2 14 31 20

Category B

Adria, Praha 1, Václavské náměstí 26, Tel. 36 04 72
Albatros (Botel), Praha 1, nábřeží Ludvíka Svobody, Tel. 2 31 36 34
Atlantic, Praha 1, Na poříčí 9, Tel. 2 31 85 12

Erko, Praha 9, Kbely 723, Tel. 89 21 05
Hybernia, Praha 1, Hybernská 24, Tel. 22 04 31
Juventus, Praha 2, Blanická 10, Tel. 25 51 51
Koruna, Praha 1, Opatovická 16, Tel. 29 39 33
Kriván, Praha 2, Náměstí I. P. Pavlova 5, Tel. 29 33 41–44
Lunik, Praha 2, Londýnská 50, Tel. 25 27 01
Merkur, Praha 1, Těšnov 9, Tel. 2 31 69 51
Meteor, Praha 1, Hybernská 6, Tel. 22 92 41, 22 42 02
Michle, Praha 4, Nuselská 124, Tel. 42 71 17
Modrá Hvězda, Praha 9, Jandova 3, Tel. 83 02 91
Moráň, Praha 2, Na Moráňi 15, Tel. 29 42 53, 29 42 51
Opera, Praha 1, Těšnov 13, Tel. 2 31 56 09
Praga, Praha 5, Plzeňská 29, Tel. 54 87 41
Racek, Praha 4, Dvorecká Louka, Tel. 42 60 51, 42 57 93
Savoy, Praha 1, Keplerova 6, Tel. 53 74 57
Transit (Interhotel), Praha 6, Ruzyně, 25. února 197, Tel. 36 71 08
U tří pštrosů, Praha 1, Malá Strana, Dražického náměstí 12,
Tel. 53 61 51
Union, Praha 2, Jaromírova 1, Tel. 43 78 58
Vltava, Praha 5, Zbraslav II, Žitavského 115, Tel. 59 15 49

Balkan, Praha 5, Třída svornosti 28, Tel. 54 07 77   Category C
Hvězda, Praha 6, Na rovni 34, Tel. 36 89 65, 36 80 37
Moravan, Praha 7, Dimitrovo náměstí 22, Tel. 80 29 05
Na Kopečku, Praha 4, Modřanská 199, Tel. 46 05 38
Nárdní dům, Praha 3, Bořivojova 53, Tel. 27 53 65
Stará Zbrojnice, Praha 1, Všehrdova 16, Tel. 53 28 15
Tichý, Praha 3, Kalininova 65, Tel. 27 30 79

Information

Čedok (Czechoslovak State Travel Bureau),   In the United Kingdom
17–18 Old Bond Street,
London W1X 4RB
Tel. 071–629 6058

Čedok,   In the United States
10 East 40th Street,
New York NY 10016
Tel. (212) 689 9720

Čedok,   In Prague
Praha 1, Na Příkopě 18, Tel. 2 12 71 11
Pražska informačni služba (Prague Information Service),
Praha 1, Na Příkopě 20, Tel. 54 44 44

The Prague Information Service publishes "A Month in Prague",
with information about events in Prague during the current
month. It also produces a 12-page brochure with the pro-
grammes of all theatres and cinemas and information about
concerts and exhibitions.

Ruzyné Airport, Tel. 36 78 02–03   Other Čedok offices
Hotel Panorama, Milevská ulice 7, Tel. 41 61 11

Čedok provides not only all the information visitors will require
but a variety of other services. It will obtain tickets for theatres,
concerts and other events, admission to night-spots, air or rail

tickets, seat reservations, shooting or fishing permits, etc.; it will arrange car hire, advise on the best garages to repair your car and change travellers' cheques; and it organises city tours and coach excursions (see Sightseeing).

Room booking service

Čedok, Praha 1, Panská 5, Tel. 22 56 57

City guides

Pragtour, Praha 1, U Obecního domu 2, Tel. 2 31 72 81
Private guides can be hired through PIS at Panská 4,
Tel. 22 34 11, 22 43 11, 22 60 67

Čedok branch for city tours, excursions and theatre tickets
Praha 1 (Staré Město), Bilkova 6 (opposite Hotel Intercontinental),
Tel. 2 31 87 69.

Language

It is not necessary to know any Czech to go to Prague. The staff of hotels, travel agencies, etc., with whom visitors will be in contact will speak some English, and young people tend to know English or French; older people may still speak some German.

It is an advantage, however, to have a few words of Czech. The pronunciation is not difficult, though the diacritic marks make it look rather fearsome.

The stress is almost invariably on the first syllable of a word: remember that the semi-vowels *l* and *r* may also carry the accent. Vowels may be either short or long. A long vowel is indicated by an accent (or in the case of *u* by a small circle). The *e* with a reverse circumflex accent (ě) is pronounced *ye*.

The vowels are pronounced in the continental fashion, without the diphthongisation found in English. Consonants are much as in English, with the following special cases:

c	= ts	ř	= rzh as in surgeon
č	= ch as in church	š	= sh as in shush
ch	= ch as in loch	ž	= zh as in treasure
ň	= ny as in canyon	dž	= j as in judge

Some useful words

address	adresa	luggage	zavazadlo
bank	banku	name	jméno
bill	účet	no	ne
bread	chléb	pay (verb)	platiti
chemist's shop	lékárna	please	prosím
doctor	lékař	post office	pošta
English	anglický	railway station	nádraží
good	dobrý	thank you	děkuji
help	pomoc	without	bez
I	ja	yes	ano
lavatory	záchod		

Topographical terms

hill	hora	square	náměstí
chapel	kaple	palace	palác
church	kostel	street	třída, ulice
bridge	most	tower	věž
embankment	nábřeží		

1 jeden	4 čtyři	7 sedm	10 deset	Numbers
2 dva	5 pět	8 osm	20 dvacet	
3 tři	6 šest	9 devět	100 sto	

Names for the most common foods and drinks can be found under Food and drink.

Libraries

The State Library of the ČSSR comprises the following separate libraries:

University Library
National Library
Slav Library

French Library
State Library of Technology
See A to Z – Clementinum

Library of the National Museum
See A to Z – National Museum

Library of Art History and Applied Art
See A to Z – Museum of Applied Art

National Literary Memorial and Museum of Czech Literature
See Strahov Abbey

Municipal Library
(Newspapers, reading room)
Praha 1, náměstí dr. Vacka 1

Library in the House of Soviet Science and Culture
Praha 1, Rytířská 31

Lost property offices

Praha 3, Olšanská 2   Documents

Praha 1, Staré Město, Bolzanova 5, Tel. 24 84 30   Other objects

Medical Help

A visitor who becomes ill should first seek help at the hotel reception desk or from the holiday courier.

Fakultni poliklinika   Medical service for
Praha 2 (Nové Město), Karlovo náměstí 32   foreigners
Tel. 29 93 81

Motoring

General Practitioner	Mon.–Fri. 8 a.m.–4.15 p.m.
Dentist	Mon.–Fri. 8 a.m.–3 p.m.
Emergency	Tel. 155
Ambulance	Tel. 333
Emergency dental service	Praha 1 (Nové Město), Vladislavova 22 Tel. 26 13 74 Mon.–Thur. 7 p.m.–7 a.m., Fri. 7 p.m.–Mon. 7 a.m.

Motoring

Regulations	Traffic regulations are much the same in Czechoslovakia as in other European countries. Vehicles travel on the right, with overtaking (passing) on the left. The penalties for traffic offences are high.
	Safety-belts must be worn. Driving after taking alcohol is absolutely prohibited.
	It is permissible to give a lift to a hitch-hiker. Hitch-hiking is prohibited only on motorways.
	Supplementary brake lights must either be masked or disconnected.
	Accidents, however slight, must in all cases be reported to the police.
	Penalties for the infringement of traffic regulations are extremely severe.
Speed limits	On motorways: 110 km/68 miles per hour Outside built-up areas: 90 km/56 miles per hour In built-up areas: 60 km/37 miles per hour
	Motorcycles and vehicles towing a trailer are restricted to 80 km/50 miles per hour outside built-up areas.
	At level-crossings speed must be restricted to a maximum of 30 km/18 miles per hour for a distance of 30 m/33 yd before reaching the crossing until the vehicle has passed over the tracks.
Petrol (gasoline)	Tusex petrol coupons at reduced prices (for "Special" – 90 octane and "Super" – 96 octane) can be obtained from Čedok. Unleaded fuel (91 octane) can be purchased using coupons for Super. Coupons for diesel – available only at certain filling stations – are issued by Čedok and Alimex representatives. In Czechoslovakia all these coupons can be obtained at the frontiers, at state banks

and Čedok tourist offices. A 20-litre reserve fuel can may be imported without payment of duty.

The following petrol stations selling super grade petrol are always open:
Praha 3 (Žižkov), Kalšnická
Praha 3 (Žižkov), Olšanská
Praha 4 (Újezd, Průhonice), on E 14 (motorway to Brno)
Praha 5 (Motol), Plzenská-Podháj
Praha 6 (Vokovice), Leninova (direction of airport)
Praha 8 (Karlín), Karlínské náměstí
Praha 9 (Hrdlöřezy), Českobrodská
Praha 9 (Prosek)

The city centre of Prague is divided into three parking zones, A, B, and C. In these zones parking is prohibited, apart from at a few parking places with progressive fees. Wenceslas Square and the adjoining streets are barred to traffic, except for buses, delivery vehicles and private cars belonging to guests staying at the hotels around the square. Motorists are advised to park either on the larger waiting areas on the Náměstí Gorkého and the main station, or at one of the car parks outside the city.
A distance of at least 3.5 m/12 ft must be maintained between a parked vehicle and tram tracks.
Parking in "no parking" areas should be avoided. The police are tough with offenders – even with foreigners – and are quick to tow away vehicles parked in the wrong place. The pound for cars so removed is in Černokostelecká nám, Praha 10 (Hostivař); Tel. 6 01 44. The charge for releasing an impounded vehicle is considerable.

Parking

Visitors arriving by car should ask at the frontier for a list of the telephone numbers of the breakdown service (Silniči Služba), the "Yellow Angels", whose patrols operate on all main roads in Czechoslovakia and in case of emergency can be summoned by telephone. In Prague itself breakdown assistance and tow-away service can be called at 22 49 06 and 22 35 44–9.

Breakdown assistance

The Autoturist office at Opletalova 29, Praha 1, Tel. 23 35 44–9, supplies information and advice, issues up-to-date information about road conditions (road maps, town plans), and can arrange accommodation in hotels and motels and on camping sites.

Autoturist

Tel. 24 24 24.

Traffic police

Praha 4 (Spořílov), Severní XI, Tel. 76 67 51–53; reception, Tel. 22 61 96.

Repairs

Praha 1 (Nové Město), Opletalova 9;
Sales office, Václavské náměstí 18, Tel. 29 05 13.

Garage service

See Travel Documents

Documents

Museums (See also Art Galleries)

Alois Jirásek and Mikoláš Aleš Museum
See A to Z – White Mountain, Star Castle

Dvořák Museum
See A to Z, Villa Amerika

Museums

Ethnographic Museum
See A to Z

Gottwald Museum (Exhibition on the Czech workers' movement
and the Czech Communist Party)
Praha 1 (Staré Město), Rytířská 29
Open Tues.–Sat. 9 a.m.–5 p.m., Sun. 9 a.m.–3 p.m.

Historical Museum
See A to Z – National Museum

Hrdlička Museum (Anthropological collection)
Praha 2, Viničná 7
Open only by appointment (Tel. 29 79 41)

Jewish Museum
See A to Z – Josefov

Komenský Museum
See A to Z – Waldstein Palace

Lenin Museum
Praha 1 (Nové Město), Hybernská 7
Open Tues.–Sat. 9 a.m.–5 p.m., Sun. 9 a.m.–3 p.m.

Military Museum of the Czechoslovak Army
Praha 3 (Žižkov), U Památníku 2
Open Nov.–March; Mon–Fri. 8.30 a.m.–5 p.m.
April–Oct.: Tues.–Sun. 9.30 a.m.–4.30 p.m.

Schloss Sbraslav

Mozart Museum
See A to Z – Bertramka

Municipal Museum
See A to Z

Museum of Applied Art
See A to Z – St Agnes's Convent and Museum of Applied Art

Museum of the Corps of National Security and Troops of the
Ministry of the Interior (Police Museum)
Praha 2, Ke Karlovu 453
Sept.–Jun: Tues.–Sun. 10 a.m.- 5 p.m.

Museum of Czech Literature
See A to Z – Strahov Abbey

Museum of Military History
See A to Z – Hradčany Square, Schwarzenberg Palace

Museum of Natural History
See A to Z – National Museum

Museum of Physical Education and Sport
See A to Z – Tyrš House

Museum of Postal and Telecommunications Services
To be re-opened in the 1990s in Vávrův dům, Praha 1,
Novomlýnska

Museum of Musical Instruments
See A to Z – Grand Prior's Palace

Náprstek Museum (Ethnological Collection) of Asiatic, African
and American cultures)
Praha 1 (Staré Město), Betlémské náměstí 1
At present under restoration

National Museum
See A to Z

National Museum of Technology
See A to Z

Smetana Museum
See A to Z – Smetanakai

Music

National Theatre (Národní divadlo)   Classical music
Praha 1 (New Town), Národní třída 2
Opera, Operetta and ballet

Smetana Theatre (Smetanovo divadlo)
Praha 1 (New Town), Vítězného února 6
Opera and ballet

Tyl Theatre (Tylovo divadlo)
Praha 1 (Old Town), Železná ulice 11
Closed until 1991 for restoration

Music Theatre (Divadlo hudby)
Praha 1 (New Town), Opletalova 5
Opera, ballet, operettas
(Large exhibition hall and extensive collection of recordings)

Karlin Music Theatre (Hudební divadlo v Karlině)
Praha 8 (Karlin), Krizikova 10.
Operettas and musicals

Concerts

With their excellent acoustics, Prague's churches are frequently used for concerts and recitals: St Vitus's Cathedral (see Hradčany), St Nicholas's (see Lesser Quarter Square), St James's, etc.

Concert halls

St Agnes's Convent (Josef Mánes Hall) Praha 1, ulice U milosrdných 17
Artists' House (Dvořák Hall), Praha 1, Nám krasnoarmejců
Hotel Intercontinental (Congress Hall), Praha 1, Nám Curieových
Palace of Culture, Praha 4, ulice 5, května 65
Lobkowitz Palace, Praha 1, Hradčany
Martinitz Palace, Praha 1, Hradčanské nám 67–68
Waldstein Palace (Knights' Hall), Praha 1, Valdštejnské nám.
Representation House (Smetana Hall), Praha 1, Nám. Republiky
Bertramka Villa, Praha 5, Mozartova 115

Prague Spring

The Prague Spring Musical Festival, held in June, has established an international reputation. During the festival concerts are given in the Palace of Culture, in churches, in various historical buildings and some in the Baroque gardens of Prague.

Musical cabaret

Rocoko Theater (Divadlo Rokoko)
Praha 1 (Nové Město), Václavské nám. 38

Jazz

Pražan, Julius Fučík Culture and Recreation Park, Praha 7 (Holešovice)
Reduta, Praha 1, (Nové Město), Národní tř 20
Metro, Praha 1, (Nové Město), Národní tř 25

Rock and Pop concerts

Palace of Culture, Praha 4, ulice 5, května 65
Julius Fučík Culture and Recreation Park, Praha 7, U Sjezdového paláce
Lucerna, Praha 1, Vodičkova 36

Night Life

Compared to other European capitals night-life in Prague is relatively minor. Concerts, and theatre and cinema performances usually finish by 10 p.m.; beer bars normally close at 11 p.m. and wine bars by midnight. Those who prefer later entertainment have a choice of jazz cellars, nightclubs in the larger hotels and a few discothèques.

Nightclubs

Alfa, Praha 1, Václavské nám 28
Café, with dancing to a band; open daily 6 p.m.–1 a.m.

Alhambra, in the Ambassador Hotel, Praha 1, Václavské nám. 5
Black theatre, revues and variety programme

Barberina, Praha 1, Melantrichova 10
Wine bar with music for dancing
Open: Mon.–Sat. 8 p.m.–4 a.m.

Embassy Club in the Ambassador Hotel, Praha 1,
Václavské nám. 5
Variety programme

Galaxie Nightclub of the Forum Hotel, Praha 4, Kongresová ulicé;
daily 9 p.m.–4 p.m.

Interconti-club of the Intercontinental Hotel
Praha 1, Nám., Curieových
Revue and cabaret

International Club of the International Hotel
Praha 6, Nám., Družby 1
Evening entertainment with folk dancing, wind band and folklore
groups

Lucerna-Bar, Praha 1, Vodičkova 36
Cabaret, variety and dancing
Daily 8 p.m.–3 a.m.

There are other nightclubs in the Hotels Central, Esplanade, Jalta,
Park and Tatran

See Music, Jazz   Jazz cellars

All near Wenceslas Square   Discothèques
Adria, Národní tř. 40: from 8.30 p.m.
Astra, Václavské nám. 28: daily 7 p.m.–2 a.m.
Habanna, V jamé 8: daily 9.30 p.m.–4.30 a.m.
Rostov, Václavaské nám. 21; daily 8 p.m.–3 a.m.
Video-Disco in the Hotel Zlatá Husá, Václavské nám. 7;
daily 7.30 p.m.–2 a.m.

Opening times

Most food shops are open on weekdays from 8 or 9.30 a.m. and   Shops
close at 6 p.m. (noon on Sat.). Small shops generally close for two
hours between noon and 3 p.m. A few large stores are open
continuously until 8 p.m. (Sat. until 4 p.m.).

The Czechoslovakian State Bank and the Živnostenská Bank in   Banks
the city centre are open Mon.–Fri. 8 a.m.–5 p.m. (Sat. 8.30 a.m.–
1 p.m.).
Other banks in Prague are normally open Mon.–Fri. 8 a.m. until
noon or 2 p.m.

Open Mon.–Fri. 8.30 a.m.–5 p.m.   Public offices

Open Tues.–Sun. 10 a.m.–5 or 6 p.m.   Galleries

Open Tues.–Sun. 10 a.m.–5 p.m. (but for the opening times of the Museums
various museums see A to Z, also Museums). The Jewish
Museum is closed on Saturday but open on Monday.

Postal services

Castles and palaces	Open Tues.–Sun. 10 a.m.–5 p.m.: from November to March they are normally closed (also on the day following official public holidays). There are exceptions to the times given above.
Čedok offices	Open Mon.–Fri. 8 a.m.–4.15 p.m.; Sat. 8.15 a.m.–noon

Postal services

Post offices	The main post office (Praha 1 (Nové Město), Jindřišská 14; tel. 26 48 41) is open day and night. Other post offices are open: Mon.–Fri. 8 a.m.–7 p.m.; Sat. 8 a.m.– noon. Some small branch post offices open Mon.–Fri. 8 a.m.–1 p.m.
Stamps	Stamps (znamka) can be obtained from post offices and tobacconists
Postage	Within Czechoslavakia: letters (dopsis) 1 Kčs; postcards 0.50 Kčs.
	To all countries in western Europe: letters (up to 10 gr) 4 Kčs, (up to 50 gr) 7 Kčs; postcards 3 Kčs.
	To all other countries: letters (up to 10 gr) 6 Kčs, (up to 20 gr) 9.6 Kčs, (up to 30 gr) 12.6 Kčs

Post box

	There are supplements for registered mail (diporučeňe) and for express delivery.
Postal code	Prague ČSSR-11000 (varies for different districts)

Public holidays and commemoration days

January 1 (New Year's Day), Easter Monday, May 1 (Labour Day), May 9 (National Day), October 28 (declaration of independence), December 25 and 26 (Christmas)

Public transport

Trams, buses and metro	Flat-rate fare 1 Kčs. On the trams a new ticket must be used whenever the passenger changes. Tickets can be bought in hotels and railway stations, at newspaper and tobacco kiosks and at transport offices.
	Tickets must be bought in advance and must be punched by the passenger in the cancelling machine on entering the vehicle. There are no conductors.
	On the principal routes there are frequent services throughout the day (during the night at intervals of about 40 minutes).
Underground (Metro)	For the Metro (see plan at end of book) the same tickets are used as for trams (see above); they are cancelled on entering the

station. On the metro passengers can change trains as often as they like; tickets are valid for one hour. The trains run from 5 a.m. until shortly before midnight. Children below 10 and senior citizens over 70 travel free.

Railway stations

Prague has some 40 railway stations. Trains from West Germany and Austria arrive at the Central Station, trains from Berlin at the Střed Station.

Central Station (Hlavní nádraží)
Praha 2 (Nové Město), Třída Vítězného února

Střed Station (Praha Střed)
Praha 1 (Nové Město), Hybernská ulice 13
This was the station from which the first train ran from Prague to Vienna in 1845.

ČSD, Tel. 2 44 44 41–9 and 26 49 30   Train information

Tickets are obtainable at the stations or from the Čedok branch at   Tickets
Na Příkopě 18 (see information).

Restaurants

Restaurants are classed in four categories according to quality, amenity and price. The category of a restaurant is shown by a Roman figure on the menu.

Bohemian specialties: see Food and drink

Alcron, Praha 1, Štěpánská 40   **Restaurants with**
Ambassador, Praha 1, Václavské náměstí 5   **international cuisine**
Barrandov, Praha 5, Kříženeckého náměstí 322
Esplanade, Praha 1, Washingtonova 19
*Evropa, Praha 1, Václavské nam. 29
   Art Nouveau restaurant with French cuisine
Flora, Praha 3, Vinohradská 121
Forum, Harmonie Restaurant, Praha 4, Kongresová ulice
*Zlatá Praha, on top floor of Intercontinental Hotel, Praha 1,
   náměstí Curieových (reservation essential)
Intercontinental (ground floor), Praha 1, náměstí Curieových
International, Praha 6, náměstí Družby 1
Olympik, Praha 8, Invalidovna, Sokolovská tř. 138
Oživlé dřevo, Praha 1, Strahovské nádvoří
Palace, Praha 1, Panská 12
Pelikan, Praha 1, Na Příkopě 7
Praha Expo 58, Praha 7, Letenské sady
Savarin, Praha 1, Na Příkopě 10
Vikárka, Praha 1 (Hradčany), Vikářská 6
Vysočina, Praha 1, Národní 28

Baltic Grill (Rybárna), Praha 1, Václavské náměstí 43 (fish and   **Speciality restaurants**
poultry)   Fish, game and poultry

Restaurants

Fregata, Praha 2 (Vyšehrad), Ladova 3 (fish)
Halali grill (Rybárna), Praha 1, Václavské náměstí 43 (fish and poultry)
Myslivna, Praha 3, Jagellonská 21 (game)
Rybárna, Praha 1, Václavské náměstí 43 (fish)
Vikárka, Praha 1 (castle), Vikářská 6 (game)
Valdštejnská Hospoda, Praha 1, Valdštejnské náměstí 7 (fish, game and poultry)

Bohemian cuisine	Chodské pohostinství, Praha 5, Na újezdé 5 Bohemian restaurant, Hotel Forum, Praha 4, Kongreslová Pálava, Praha 3, Slavíkova 18 Staropražská Rychta, Praha 1, Václavskáe náměstí 7 Šumavan, Praha 3, Ondříčkova 17 V sloupu, Praha 3, Lucemburská
Bulgarian cuisine	Sofia, Praha 1, Václavské náměstí 33
Chinese cuisine	Babetka, Praha 3, Ževlivského 10 Čínská restaurace, Praha 1, Vodičkova 19
German cuisine	Alex, Praha 1, Revolučni 11 (in style of Berlin in the 1930s)
Italian cuisine	Trottaria Viola, Praha 1, Národní tř. 7
Indian cuisine	Indická restaurace Mayur, Praha 1, Stepanská 60
Jewish cuisine (kosher)	Restaurant in the Jewish Town Hall, Praha 1, Maislova 18
Russian cuisine	Berjozka, Praha 1, Rytířska 7 Gruzia, Praha 1, Na příkopě 29
Slovak cuisine	Paříž, Praha 1, V Úbecního domu 1
Yugoslav cuisine	Beograd, Praha 1, Vodičova 5 Jadran, Praha 1, Mostecká 21
Dietary restaurants	Akademická, Praha 1, Vodičkova 15 Dieta, Praha 2, Francouzská 2 Petrská dietni restaurace, Praha 1, Petrské náměstí 1 Regent Praha, Praha 1, Karmelitská 20
Garden and terrace restaurants	Barrandov, Praha 5, Kříženeckého náměstí 322 Savarin, Praha 1, Na příkope 10 Slovanský ostrov, Praha 1 (island near the National Theatre) Plzeňský dvůr, Praha 7, Úbrránců miru 59 Slovanská hospoda, Praha 1, Na Příkopě 22 U Fleků (see Beer Parlours) Praha-Expo 58 (see Restaurants with international cuisine)
Beer-parlours	See Practical Information: Beer
Wine-bars	Fregata, Praha 2, Ladova 3 Intercontinental, Praha 1, náměstí Curieových Klášterní vinárna, Praha 1, Národní 8 Lobkovická vinárna, Praha 1, Vlašská 17

Lví Dvůr, Praha 1 (Hradčany), U Prasného mostu 6
Makarská vinárna, Praha 1, Malostranské náměstí 2
Mělniká vinárna, Praha 1, Národní tř. 17
Obecní dům, Praha 1, náměstí Republiky 1090
Opera-Grill, Praha 1, Divadelní 24
Slovenská vinárna, Praha 1, Františkánská zahrada
U Fausta, Praha 2, Karlova nám. 4
U Golema, Praha 1, Maislova 8
U labutí, Praha 1, Hradčanské náměstí 11
U malířů, Praha 1, Maltézské náměstí 11
U markýze, Praha 1, Nekázanka 8
U mecanašé (see Restaurants with international
  cuisine)
U palcátu, Praha 1, Thunovská 16
U patrona, Praha 1, Dražického náměstí 4
U pavouka, Praha 1, Celetná ulice 17
U plebána, Praha 1, Betlémské náměstí 10
U Rudolfa, Praha 1, Maislova 5
U Šuterů, Praha 1, Paleckého ulice 4
U tři housliček, Praha 1, Nerudova 12
U zelené žáby, Praha 1, U radnice 8
U zlaté hrušky, Praha 1, Novy svět 3
U zlatého jelena, Praha 1, Celetná 11
*U zlaté konvice, Praha 1, Melantrichova 20
  (gipsy music in the Gothic vaults)
U zlaté štiky, Praha 1, Dlouhá 9
*Viola, Praha 1, Národní tř. 7
  (first-class poetry evenings are held in the little theatre of this
  wine-bar)
Vinný sklípek pana Broučka, Praha 1 (Hradčany)

Shopping

The State-run Tuzex shops sell goods only for convertible foreign
currency or "Tuzex crowns", which can be bought in Czechoslovak banks at the official rate of exchange.

Tuzex

Visitors should keep all receipts for purchases in Tuzex shops.

Information on duty-free export of goods purchased: Tuzex,
Praha 1 (Staré Město), Rytířská 13.

When buying crystal, *objets d'art*, etc., visitors should seek
advice from Tuzex staff, since some articles of this kind are
subject to heavy export duties. Tuzex takes care of all formalities
and arranges dispatch.

Antiques: Staré Město, Rytířská 43
Glass, china, perfume, souvenirs: Staré Město, Železná 18
Perfume: Nové Město, Smečky 23
Textiles, ladies' underwear: Nové Město, Spálená 4
Ladies' and children's clothing: Praha 7, Jancovcova 2
Furs, clothing: Praha 1, Lazarská 82
Musical Instruments: Praha 1 (Nové Město). Jungmannovo
náměstí 17

Tuzex shops

See Antiques

Antiques

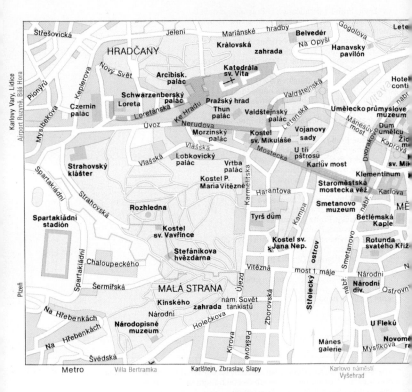

Metro   Villa Bertramka   Karlštejn, Zbraslav, Slapy   Karlovo náměstí
Vyšehrad

Art (Dílo galleries)	Nové Město, Vodičkova 32 Platýz, Staré Město, Narodní 37 Karolina, Staré Město, Železná 6 Centrum, Praha 1 (Staré Město) 28, října 6 Zlatá lilie, Staré Město, Malé náměstí 12 Praha 1 (Hradčany), Zlatá ulička (Golden Lane) Praha 4 (Nusle), Nuselská 5
Arts and crafts and folk art	All in Praha 1 Česká jizba (Czech room): Staré Město, Karlova 12 Slovenská jizba (Slovak room): Nové Město, Václavské nám. 40 ÚLUV (headquarters for folk art): Staré Město, Národní třída 36 ÚVA: Staré Město, Na Příkopě 25–27 Dielo-SFVU (Slovak fund for fine art): Staré Město, Na Příkopě 5 Výtvarna řemesla (craftwork): Staré Město, Křížovnické nám. 2
Books	Kniha, Praha 1, Štepánská 42 (literature in foreign languages) Kniha, Praha 1, Vodičkova 21 (particularly illustrated books) Kniha, Praha 1 (Staré Město), Karlova 14 (graphic art, etc.) Československý spisovatel, Praha 1, Národní 9 Knihkupectví Melantrich, Praha 1, Na Příkopě 3–5 Knihkupectví U zlatého klasu, Praha 1, Na Příkopě 23

Recommended city tour

All in Praha 1   Glass and porcelain

Bohemian Glass: Staré Město, Pařížká 2
Bohemian Moser: Staré Město, Na Příkopě 12
Crystalex: Staré Město, Malé nám. 6 (glass from Bor-Haida)
Diamant: Nové Město, Václavské nám. 3
Krystal: Nové Město, Václavské nám. 30
Krystal: Staré Město, Celetná 20

All in Praha 1:   Jewellery

Bijoux de Bohême, Staroměstské náměstí 6
Na Příkopě 12
Václavské náměstí 53
Václavské náměstí 28 (garnets)
Vodičkova 39
Národní třída 25
28. října 3
Obchodní dům Družba, Václavské náměstí 21

All in Praha 1   Records (Supraphon)

Malá Strana, Mostecká 9
Staré Město, Celetná 8

Sightseeing

Nové Město, Jungmannova 20
Nové Město, Jindřišská 19
Nové Město, Václavské náměstí 17
Nové Město, Václavské náměstí 51

Sports articles   All in Praha 1

Dům sportu (House of Sport): Nové Město, Jungmannova 28
Tyršův dům: Malá Strana, Újezd 42
Sportovní potřeby: Nové Město, Vodičkova 30

See also Department Stores

Sightseeing (See map pages 208/9)

City tours   Čedok (see Information) organises city tours to the most attractive places in the "Golden City", with commentary in English.

A three-hour tour of the city "Historic Prague" partly by bus and partly on foot takes place daily throughout the year. A one-hour tour ("Prague by Night"), with a subsequent folklore programme, operates from mid-May to mid-October every Wednesday and Friday. Included are dinner and a visit to a Bohemian wine-bar.

From May to October on Tuesdays and Thursdays there is a three-hour tour on foot through the historic Old Town, then by cable-car up Petřín Hill and continuing on foot to Strahov Convent, the Loreto Shrine and Prague Castle.

Boat excursions on the Vltava   See entry

Old-timer   Three-hour tour by old-timer tram through the Old Town and the Lesser Quarter (July–October).

Excursions   For excursions into the surroundings of Prague see entry.

Sports

Baths (covered)   Podolí, Praha 4, Podolská 74
Klárov, Praha 1, nábřeží kapitána Jaroše

Baths (open-air)   Občanská plovárna, Praha 1, nábřeží kapitána Jaroše
Ljotka, Praha 4, Novodvorska
Štvanice, Praha 7, Holešovice, ostrov Štvanice Island

Football   Stadion Bohemians ČKD, Praha 10, Vršovice SNB 2
Stadion Dukla, Praha 6 (Dejvice), Na Julisce 28 (army football)
SK Slavja Praha IPS, Praha 10, Vršovice, Stadion dr. V. Vacka
Sparta ČKD Praha, Praha 7, Letná, Obránců minu 98

Golf   TJ Golf
Praha 5-Motol

Horse-racing   State racecourse, Chuchle

Tennis   Sparta ČKD Praha, Praha 7, Stromovka
Slavia Praha IPS, Praha 7, Letná Kostelní
Dopravní podnik, Praha 7, Štvanice

Taxis

Taxis can be called by telephone, hired at taxi ranks or hailed in
the street.

Tel. 20 29 51 and 20 39 41   24-hour service

Telephone

Within the city local calls cost 1 Kčs and are not timed. When   Local calls
using a call box the crown is put in the slot as soon as the number
called answers. The cost of a three-minute call within Czechoslovakia is between 5 and 15 Kčs
Call-boxes in Prague are not always in working order.

Calls to other countries are best made from an hotel or a post-   Trunk calls
office by prior arrangement, as calling from a phone-box entails
putting in 5-crown coins every time a signal is heard. The cost of
calls to other countries is relatively high.

From Prague you dial 00, then the code for the country concerned   Self-dialling
followed by the local code (omitting the 0) and finally the number
of the subscriber.

From the United Kingdom the dialling code for Prague is: Codes
010 42 2; from the USA and Canada 011 42 2.

To the UK from Prague 00 44; to the USA and Canada 001

Theatres

Since performances are usually sold out tickets should be bought   Tickets
in advance (see Information – Čedok). General booking in
advance (Mon.–Fri. 10 a.m.–6 p.m.)

SLUNA, Praha 1, Panská 4, pasáž Černá růže; tel. 22 12 06

SLUNA, Praha 1, Václavské nám. 28, pasáž Alfa; tel. 26 16 06,
16 06 93

Divadlo ABC (ABC Theatre)   Theatres
Praha 1 (Nové Město), Vodičkova 28
Comedies

Atelier Ypsilon
Praha 1 (Nové Město), Spálená 16
Music and text

Braniké Divadlo (Branik Theatre)
Praha 4 (Nusle), Braniká 63
Modern pantomime

Divadlo E. F. Buriana (Burian Theatre)
Praha 1 (Nové Město), Na poříčí 26
Formerly an avant-garde theatre

DISK
Praha 1 (Staré Město), Karlova 8
Young actors and actresses, classical repertoire
Some performance in the Palace of Culture

Laterna Magica
Praha 1 (Nové Město), Národní třída 40
A combination of music, drama, dancing, miming and film

Palác kultury (Palace of Culture)
Praha 4 (Nusle) ul. 5, května 65
Mixed programme

Lyra Pragensis
Praha 1 (Hradčany), Hradčanské náměstí 8
Cabaret

Maltese Garden
Praha 1 (Malá Strana) Velkopřevorské nám.
Plays

Národní divadlo (National Theatre)
Praha 1 (Nové Město), Národní třída 2
Opera, ballet, drama

Nová scéna (New Scene)
Praha 1 (Nové Město), Národní třída 4
Ballet and plays

Divadlo S. K. Neumanna (Neumann Theatre)
Praha 8 (Libeň), Rudé armády 34
A popular suburban theatre

Realistické divadlo (Realist Theatre)
Praha 5 (Smíchov), ulice S. M. Kirova 57
Socialist realism, topical plays

Reduta
Praha 1 (Nové Město), Národní třída 10
Little theatre

Činoherní klub (Drama Club)
Praha 1 (Nové Město), Ve smečkách 40

Černé divadlo (Black Theatre)
No permanent company, generally touring abroad

Divadlo Semafor (Semaphore Theatre)
Praha 1 (Nové Město), Václavské náměstí 28
Musical comedy, mime

Smetana Theatre
Praha 1 (Nové Město), Vítězného února 8
Opera, ballet and plays

Smetana Theatre: the foyer

Divadlo Na zábradlí (Theatre at the Railings)
Praha 1 (Staré Město), Anenské náměstí 5
Following the tradition of Jean Gaspard Debureau, born 1796 in
Kolín, of the immortal pierrot-pantomime of the Théâtre des
Funambules in Paris and of the famous mime artists Louis Bar-
rault and Marcel Marceau, Ladislav Fialka continues to present
classical pantomime, complemented by new modern forms.

Tylovo divadlo (Tyl Theatre)
Praha 1 (Staré Město), Železná ulice 11

Divadlo na Vinohradech (Vinohrady Theatre)
Praha 2 (Vinohrady), náměstí Míru 7
Classical and contemporary plays

Juniorklub Na chmelnici
Praha 3 (Žiškov), Konévova 219
Music, theatricals, cabaret

Kabarett u Fleků   Cabaret
Praha 1 (Nové Město), Křemencova 11

Divadlo Rokoko (Rococo Theatre)
Praha 1 (Nové Město), Václavské náměstí 38

Variété Praga
Praha 1 (Nové Město), Vodičkova 30

Viola Wine-bar
Praha 1 (Staré Město), Národní tř. 7
Poetry readings

White Horse (Bilý koníček)
Praha 1 (Staré Město), Staroměstské nám. 20
Cabaret

Malostranské beseda
Praha 1 (Malá Strana), Malostranské nám. 21
Jazz and theatricals

Children's and young people's theatres

Albatros
Praha 1 (Staré Město), Na Perštýně 1

Loutka Puppet Theatre
Praha 2 (Nové Město), náměstí Maxima Gorkého 28

Říše loutek (Puppet Kingdom)
Praha 1 (Staré Město), Žatecká ulice

Slunicko
Praha 1 (Staré Město), Na Příkopě 15

Divadlo Spejbla a Hurvínka (Spejbl and Hurvinek)
Praha 2 (Vinohrady), Římská 45
Puppets

Divadlo Jiřího Wolkera (Jiří Wolker Theatre)
Praha 1 (Staré Město), Dlouhá třída 39

Time

Czechoslovakia observes Central European Time, one hour ahead of Greenwich Mean Time, seven hours ahead of Eastern Standard Time.

From April to September Summer Time (Daylight Saving Time), one hour ahead of Central European Time, is in force.

Tipping

Although a service charge is included in bills, a tip of 5–10 per cent is never unwelcome.

Travel documents

Passport

Every visitor must be in possession of a full passport, valid for at least four months at the time of application for a visa. The photograph must be a faithful likeness of the holder.

MERIDIANUS
QUO OLIM TEMPUS PRAGENSE
DIRIGEBATUR

POLEDNÍK,
PODLE NĚHOŽ BYL V MINULOSTI
ŘÍZEN PRAŽSKÝ ČAS

Historic meridian in Old Town Square

All visitors must have a visa. It is best to leave the obtaining of a visa to a travel bureau, as individual applications take a long time. Application forms are obtainable from the consular offices of the ČSSR, from major travel bureaux or from Čedok (see Information). Two passport-size photographs are required.

Applications for visas take about two weeks to process. In urgent cases the Czech consulates will issue a visa within two days.

From January 1989 visas can be obtained at frontier crossings between 7 a.m. and 7 p.m. providing the visitor enters the country on the same day. A visa will only be issued provided that the visitor complies with the compulsory currency exchange (see Currency). Transit visas (valid 48 hours) are not subject to the compulsory currency exchange.

A visitor staying at a private address must report to the police (Population Registration Headquarters; Praha 3, Olsanská 2) within 48 hours of arrival.

Driving licence, car registration and international insurance (Green Card) documents are required. If the driver does not own the vehicle the written authority of the owner must be produced.

Visa

Private visits

Car documents

When to Go

In spring, that is from about mid-April, flowering fruit trees on the slopes bordering the Vltava are a delightful feature of Prague. An additional attraction of this season are the traditional annual music weeks ("Prague Spring" – see Events). The highest temperatures are in July, a month which also has the highest rainfall and which is inclined to be stormy. Perhaps the best time to visit the "Golden City" is in the autumn, when weather conditions are good. There is not too much snow in winter, the coldest month being January with an average temperature of $-4.2\ °C/24\ °F$.

Youth hostels

Youth Hostels in the accepted sense do not exist in Prague. Students below 30 years of age are granted reductions at the Juniorhotel, Praha 2, Žitná ulice 9; tel. 29 33 41.

During the summer vacation (July and August) inexpensive accommodation is available at the Koleje VŠCHT Student House (Praha 4, Jižbi město).

Information from ČSM Youth Travel Office, Praha 2, Žitná ulice 9; tel. 29 85 87.

Useful Telephone Numbers

Emergencies	
General	155
Fire	150
Police	158
Ambulance	333
Breakdown assistance	22 49 06
Dental emergency service	26 13 74
Information	
General information on Prague	54 44 44
Bus services	22 14 45
Cinema programmes	145
City tours	2 36 23 27
Theatres and concerts (afternoon)	144
Tickets (admission)	2 36 23 56, 2 36 22 97
Time	112
Train services	26 49 30, 2 44 44 41
Weather forecast	116
Embassies	
United Kingdom	53 33 47–9, 53 33 40, 53 33 70
United States	53 66 41–8
Canada	32 69 41
Airlines	
Czechoslovak Air Lines	21 46
British Airways	24 08 47–8
Lufthansa	2 31 75 51
Lost property	24 84 30
Taxis	20 29 51, 20 39 41
Telephone	
Information, Prague	120
Information, ČSSR	121
Dialling codes	
From the United Kingdom	010 42 2
From the United States or Canada	011 42 2
To the United Kingdom	00 44
To the United States or Canada	001
Telegrams	127

Index

Notes

Notes

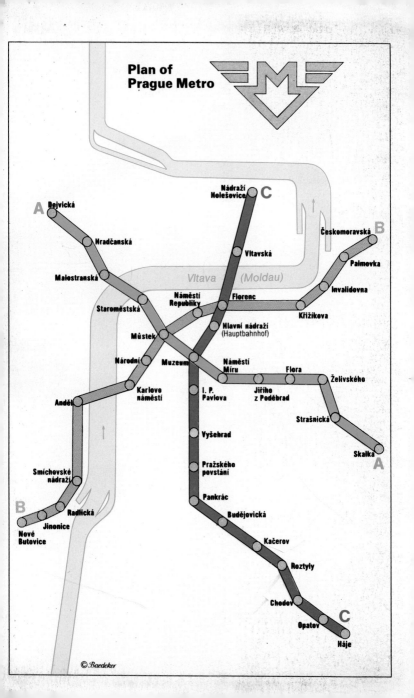